ST(P) MATHEMATICS 4A

SIMON WEBSTER

ST(P) MATHEMATICS will be completed as follows:

Published 1984
{
ST(P) 1
ST(P) 1 Teacher's Notes and Answer Book

ST(P) 2
}

Published 1985
{
ST(P) 2 Teacher's Notes and Answer Book

ST(P) 3A
ST(P) 3B
ST(P) 3A Teacher's Notes and Answer Book
ST(P) 3B Teacher's Notes and Answer Book
}

Published 1986
{
ST(P) 4A
ST(P) 4B
ST(P) 4A Teacher's Notes and Answer Book
ST(P) 4B Teacher's Notes and Answer Book
}

In preparation
{
ST(P) 5A
ST(P) 5B
ST(P) 5C
ST(P) 5A Teacher's Notes and Answer Book
ST(P) 5B Teacher's Notes and Answer Book
ST(P) 5C Teacher's Notes and Answer Book
}

ST(P)
MATHEMATICS 4A

L. Bostock, B.Sc.
formerly Senior Mathematics Lecturer, Southgate Technical College

S. Chandler, B.Sc.
formerly of the Godolphin and Latymer School

A. Shepherd, B.Sc.
Head of Mathematics, Redland High School for Girls

E. Smith, M.Sc.
Head of Mathematics, Tredegar Comprehensive School

Stanley Thornes (Publishers) Ltd

First Published 1986 by
Stanley Thornes (Publishers) Ltd,
Old Station Drive,
Leckhampton,
CHELTENHAM GL53 0DN

British Library Cataloguing in Publication Data
ST(P) mathematics
   Book 4A
   1. Mathematics—1961–
   I. Bostock, L.
   510   QA39.2

   ISBN 0–85950–250–3

Typeset by Tech-Set, Gateshead, Tyne & Wear
Printed and bound in Great Britain at The Bath Press, Avon

CONTENTS

INTRODUCTION

This book continues the work in 3A and is intended for use with pupils taking the highest level GCSE papers.

The contents of this book covers, in the main, the common syllabus requirements of all the examining boards. Other topics included by some, but not all examining boards are in 5A. Topics introduced earlier in the series are revised before being developed further.

There are plenty of straightforward questions and exercises are divided into three types.

The first type, identified by plain numbers, e.g. **12.**, are considered necessary for consolidation.

The second type, identified by a single underline, e.g. <u>**12.**</u>, are extra, but not harder, questions for extra practice or for later revision.

The third type, identified by a double underline, e.g. <u><u>**12.**</u></u>, are more demanding questions.

Multiple choice questions are included for the first time in the A series. These can provide useful self-test questions to confirm understanding. They also provide the basis for understanding class discussions. At the end of the book there are some general revision exercises of examination type questions.

ACKNOWLEDGEMENTS

The authors and publishers would like to thank the following for permission to include material.

The Financial Times: for the Exchange Cross Rates table on page 291

British Rail: for the timetable on page 379

1 ALGEBRAIC FRACTIONS

ADDITION AND SUBTRACTION

In Book 3A, Chapter 23, we simplified algebraic fractions of the form

$$\frac{x}{4} - \frac{x+3}{3}$$

In this chapter we revise that work and extend it.

EXERCISE 1a

Simplify $\dfrac{2x-1}{5} - \dfrac{x+2}{4}$

$$\frac{(2x-1)}{5} - \frac{(x+2)}{4} = \frac{4(2x-1) - 5(x+2)}{20}$$

$$= \frac{8x - 4 - 5x - 10}{20}$$

$$= \frac{3x - 14}{20}$$

Simplify:

1. $\dfrac{x+1}{3} + \dfrac{x+2}{4}$

2. $\dfrac{2x-1}{4} + \dfrac{x-3}{5}$

3. $\dfrac{4x+3}{5} + \dfrac{x-2}{2}$

4. $\dfrac{x-2}{6} - \dfrac{x-3}{7}$

5. $\dfrac{3x+2}{3} - \dfrac{2x-1}{4}$

6. $\dfrac{x+3}{2} + \dfrac{x+1}{5}$

7. $\dfrac{x+1}{3} + \dfrac{x-2}{4}$

8. $\dfrac{2x-3}{5} + \dfrac{3x+1}{2}$

9. $\dfrac{x+2}{3} - \dfrac{x-2}{4}$

10. $\dfrac{5x-2}{6} - \dfrac{3x-2}{5}$

11. $\dfrac{2x+3}{3} + \dfrac{3x-4}{6}$

12. $\dfrac{5x-2}{12} - \dfrac{x-5}{4}$

1

Simplify $\dfrac{5}{2x+3} - \dfrac{2}{5x}$

$$\dfrac{5}{2x+3} - \dfrac{2}{5x} = \dfrac{5(5x) - 2(2x+3)}{5x(2x+3)}$$

$$= \dfrac{25x - 4x - 6}{5x(2x+3)}$$

$$= \dfrac{21x - 6}{5x(2x+3)}$$

$$= \dfrac{3(7x-2)}{5x(2x+3)}$$

Simplify:

13. $\dfrac{2}{x} + \dfrac{5}{4x}$

16. $\dfrac{4}{x-4} - \dfrac{2}{x+3}$

19. $\dfrac{3}{4a} - \dfrac{2}{3a}$

14. $\dfrac{1}{5a} - \dfrac{2}{13a}$

17. $\dfrac{3}{x+2} + \dfrac{5}{2(x+2)}$

20. $\dfrac{5}{x+3} + \dfrac{2}{x-1}$

15. $\dfrac{2}{x+3} + \dfrac{3}{x+4}$

18. $\dfrac{3}{x} + \dfrac{4}{3x}$

21. $\dfrac{8}{x+7} - \dfrac{3}{x-4}$

Sometimes we factorise the denominators first so that we can choose the simplest common denominator.

EXERCISE 1b

Simplify $\dfrac{2}{x^2-9} - \dfrac{3}{x+3}$

$$\dfrac{2}{x^2-9} - \dfrac{3}{x+3} = \dfrac{2}{(x+3)(x-3)} - \dfrac{3}{(x+3)}$$

$$= \dfrac{2 - 3(x-3)}{(x+3)(x-3)}$$

$$= \dfrac{2 - 3x + 9}{(x+3)(x-3)}$$

$$= \dfrac{11 - 3x}{(x+3)(x-3)}$$

Simplify:

1. $\dfrac{3}{x+1} + \dfrac{2}{x^2-1}$

4. $\dfrac{4}{x-3} - \dfrac{1}{x^2-9}$

2. $\dfrac{5}{x^2-4} + \dfrac{3}{x+2}$

5. $\dfrac{2}{x+2} - \dfrac{x-4}{x^2-4}$

3. $\dfrac{3}{x^2-16} - \dfrac{4}{x-4}$

6. $\dfrac{7}{x^2-1} + \dfrac{2}{x-1}$

7. $\dfrac{3}{x+5} - \dfrac{2}{x^2-25}$

10. $\dfrac{1}{2x} + \dfrac{x-3}{x^2-2x}$

8. $\dfrac{5}{x^2-49} - \dfrac{9}{x-7}$

11. $\dfrac{2}{3x} + \dfrac{x-5}{x(x+3)}$

9. $\dfrac{4}{x+4} + \dfrac{3}{x^2-16}$

12. $\dfrac{3}{x^2-9} + \dfrac{5}{x+3}$

There are times when the fraction that results from reducing several fractions to a single fraction can be simplified further because there is a factor that is common to the numerator and the denominator.

EXERCISE 1c

Reduce $\dfrac{2x}{x^2-4} - \dfrac{1}{x-2}$ to a single fraction in its lowest terms.

(The first step is to factorise the denominator.)

$$\dfrac{2x}{x^2-4} - \dfrac{1}{x-2} = \dfrac{2x}{(x+2)(x-2)} - \dfrac{1}{x-2}$$

$$= \dfrac{2x-(x+2)}{(x+2)(x-2)}$$

$$= \dfrac{2x-x-2}{(x+2)(x-2)}$$

$$= \dfrac{\cancel{(x-2)}}{(x+2)\cancel{(x-2)}}$$

$$= \dfrac{1}{x+2}$$

Simplify:

1. $\dfrac{1}{x+1} + \dfrac{2}{x^2-1}$

2. $\dfrac{1}{2+x} + \dfrac{2x}{4-x^2}$

3. $\dfrac{6}{x^2-2x-8} + \dfrac{1}{x+2}$

4. $\dfrac{3}{x^2+5x+4} + \dfrac{1}{x+4}$

5. $\dfrac{2}{x-1} + \dfrac{4}{x^2-1}$

6. $\dfrac{1}{x-1} - \dfrac{x+2}{2x^2-x-1}$

7. $\dfrac{8}{x^2-2x-15} - \dfrac{1}{x-5}$

8. $\dfrac{2}{x^2+4x+3} - \dfrac{1}{x^2+5x+6}$

9. $\dfrac{6}{x^2-9} + \dfrac{1}{x+3}$

10. $\dfrac{1}{x-3} + \dfrac{1}{x^2-7x+12}$

11. $\dfrac{3}{x^2-x-2} + \dfrac{1}{x+1}$

12. $\dfrac{1}{x^2-4x+3} + \dfrac{1}{x^2-1}$

13. $\dfrac{9}{x^2+x-2} - \dfrac{3}{x-1}$

14. $\dfrac{10}{2x^2-3x-2} - \dfrac{2}{x-2}$

15. $\dfrac{x}{x^2+6x+8} + \dfrac{1}{x+2}$

16. $\dfrac{1}{x^2+9x+20} + \dfrac{2}{x^2+6x+8}$

MIXED QUESTIONS

EXERCISE 1d

Simplify:

1. $\dfrac{3x+2}{3} + \dfrac{x+1}{4}$

2. $\dfrac{5x-3}{5} - \dfrac{3x-2}{4}$

3. $\dfrac{5}{6x} - \dfrac{2}{3x}$

4. $\dfrac{6}{x-2} - \dfrac{4}{x^2-4}$

5. $\dfrac{3}{x^2-2x-8} - \dfrac{5}{x^2-5x+4}$

6. $\dfrac{2}{x+5} + \dfrac{14}{x^2+3x-10}$

7. $\dfrac{1}{x^2-7x+12} + \dfrac{1}{x^2-5x+6}$

8. $\dfrac{1}{2x^2+3x-2} - \dfrac{1}{3x^2+7x+2}$

SOLVING EQUATIONS WITH FRACTIONS

Remember that if we alter one side of an equation we must alter the other side in the same way. If an equation contains fractions we can remove them by multiplying both sides by the LCM of the denominators.

EXERCISE 1e

Solve the equation $\dfrac{x}{2} + \dfrac{x}{8} = 10$

$$\frac{x}{2} + \frac{x}{8} = 10$$

Multiply both sides by 8 $8 \times \dfrac{x}{2} + 8 \times \dfrac{x}{8} = 8 \times 10$

$$4x + x = 80$$
$$5x = 80$$
$$x = 16$$

Solve the following equations.

1. $\dfrac{x}{6} + \dfrac{x}{3} = 3$ **4.** $\dfrac{5x}{4} - \dfrac{2x}{3} = 7$ **7.** $\dfrac{2x}{5} - \dfrac{x}{3} = \dfrac{4}{3}$

2. $\dfrac{x}{4} - \dfrac{x}{8} = 1$ **5.** $\dfrac{x}{4} + \dfrac{x}{6} = 10$ **8.** $\dfrac{3x}{2} - \dfrac{4x}{7} = 13$

3. $\dfrac{2x}{3} + \dfrac{x}{5} = \dfrac{13}{3}$ **6.** $\dfrac{x}{5} - \dfrac{x}{10} = 2$ **9.** $\dfrac{5x}{3} - \dfrac{3x}{4} = \dfrac{11}{6}$

Solve the equation $\dfrac{4}{5} - \dfrac{2}{x} = \dfrac{2}{15}$

$$\frac{4}{5} - \frac{2}{x} = \frac{2}{15}$$

Multiply both sides by $15x$

$$\overset{3}{15x} \times \frac{4}{\underset{1}{5}} - 15x \times \frac{2}{\underset{1}{x}} = \overset{1}{15}x \times \frac{2}{\underset{1}{15}}$$

$$12x - 30 = 2x$$
$$10x - 30 = 0$$
$$10x = 30$$
$$x = 3$$

Solve the following equations.

10. $\dfrac{1}{2} - \dfrac{1}{x} = \dfrac{1}{4}$

14. $\dfrac{1}{2} - \dfrac{1}{x} = \dfrac{1}{6}$

11. $\dfrac{3}{2x} + \dfrac{5}{24} = \dfrac{7}{12}$

15. $\dfrac{2}{x} + \dfrac{1}{4} = \dfrac{11}{12}$

12. $\dfrac{3}{x} - \dfrac{1}{2x} = 5$

16. $\dfrac{2}{3x} - \dfrac{1}{2x} = \dfrac{1}{2}$

13. $\dfrac{7}{4x} - \dfrac{1}{2} = \dfrac{3}{8}$

17. $\dfrac{17}{7x} - \dfrac{5}{7} = \dfrac{1}{2}$

Solve the equation $\dfrac{x-2}{5} - \dfrac{x-5}{3} = \dfrac{1}{15}$

$$\dfrac{(x-2)}{5} - \dfrac{(x-5)}{3} = \dfrac{1}{15}$$

Multiply both sides by 15

$$\dfrac{^3\cancel{15}}{1} \times \dfrac{(x-2)}{\cancel{5}_1} - \dfrac{^5\cancel{15}}{1} \times \dfrac{(x-5)}{\cancel{3}_1} = \dfrac{^1\cancel{15}}{1} \times \dfrac{1}{\cancel{15}_1}$$

$$3(x-2) - 5(x-5) = 1$$

$$3x - 6 - 5x + 25 = 1$$

$$-2x + 19 = 1$$

$$19 = 1 + 2x$$

$$18 = 2x$$

$$9 = x$$

Solve the following equations.

18. $\dfrac{x-2}{2} + \dfrac{x}{3} = \dfrac{3}{2}$

22. $\dfrac{x-1}{3} + \dfrac{x}{4} = \dfrac{5}{6}$

19. $\dfrac{x-1}{9} + \dfrac{3x-7}{4} = \dfrac{13}{18}$

23. $\dfrac{2x-1}{4} + \dfrac{x-3}{5} = \dfrac{11}{20}$

20. $\dfrac{3x-2}{5} - \dfrac{x-1}{2} = \dfrac{3}{10}$

24. $\dfrac{2x-3}{6} - \dfrac{x-3}{2} = \dfrac{1}{3}$

21. $1 - \dfrac{3x+7}{2} = \dfrac{x+4}{4}$

25. $\dfrac{1}{2} - \dfrac{x-2}{5} = \dfrac{2x-3}{10}$

Sometimes a fractional equation leads to a quadratic equation which will factorise.

EXERCISE 1f

Solve the equation $\dfrac{6}{x} + \dfrac{1}{x-5} = 2$

$$\frac{6}{x} + \frac{1}{x-5} = 2$$

Multiply both sides by $x(x-5)$

$$x(x-5)\left[\frac{6}{x} + \frac{1}{x-5}\right] = x(x-5)\times 2$$

$$x(x-5)\times\frac{6}{x} + x(x-5)\times\frac{1}{x-5} = 2x(x-5)$$

$$6(x-5) + x = 2x(x-5)$$

$$6x - 30 + x = 2x^2 - 10x$$

$$7x - 30 = 2x^2 - 10x$$

$$0 = 2x^2 - 17x + 30$$

$$\therefore \qquad 2x^2 - 17x + 30 = 0$$

$$(2x-5)(x-6) = 0$$

$$\therefore \qquad \text{either} \quad 2x - 5 = 0 \quad \text{or} \quad x - 6 = 0$$

$$2x = 5 \quad \text{or} \quad x = 6$$

$$\therefore \qquad x = 2\tfrac{1}{2} \quad \text{or} \quad x = 6$$

Solve the following equations, each of which leads to a quadratic equation that factorises.

1. $\dfrac{5}{x+3} - \dfrac{1}{x} = \dfrac{1}{2}$

2. $\dfrac{1}{x} - \dfrac{1}{x+1} = \dfrac{1}{20}$

3. $\dfrac{x+5}{x} = x - 3$

4. $\dfrac{2}{x+4} + \dfrac{3}{x} = 1$

5. $\dfrac{3}{x-1} - \dfrac{x}{3} = \dfrac{1}{2}$

6. $\dfrac{8}{x} + \dfrac{1}{x-5} = 1$

7. $x + \dfrac{3}{x+4} = 0$

9. $\dfrac{3}{x} + \dfrac{2}{x-1} = 2$

8. $\dfrac{2}{x} - \dfrac{1}{x+2} = \dfrac{1}{3}$

10. $\dfrac{5}{x+1} + \dfrac{2}{x+6} = \dfrac{6}{5}$

11. $\dfrac{2}{x+1} + \dfrac{3}{x+4} = \dfrac{2}{3}$

14. $\dfrac{1}{x-3} - \dfrac{3}{x-4} = \dfrac{1}{2}$

12. $\dfrac{4}{x-8} + \dfrac{3}{x-2} = \dfrac{1}{2}$

15. $\dfrac{3}{5-x} - \dfrac{3}{4} = \dfrac{1}{x+2}$

13. $\dfrac{3}{x+2} - \dfrac{1}{x+1} = \dfrac{5}{12}$

16. $\dfrac{x+2}{4} - \dfrac{3}{x-5} = 1\tfrac{1}{2}$

2 SETS

SET NOTATION

A *set* is a collection of things having something in common.

Things that belong to a set are called *members* or *elements*. These elements are usually separated by commas and written down between curly brackets or braces.

Instead of writing 'the set of British pop stars', we write {British pop stars}.

EXERCISE 2a

1. Use the correct set notation to write down the following sets.
a) the set of teachers in my school
b) the set of books I have read.

2. Write down two members from each of the sets given in question 1.

Describe in words the set $\{2, 4, 6, 8, 10, 12\}$

$\{2, 4, 6, 8, 10, 12\}$ = {even numbers from 2 to 12 inclusive}

---

3. Write down in words the given sets:
a) $\{1, 3, 5, 7, 9\}$
b) {Monday, Tuesday, Wednesday, Thursday, Friday}

Note that these descriptions must be very precise.
e.g. it is correct to say

$$\{1, 2, 3, 4, 5\} = \{\text{first five whole numbers}\}$$

but it is incorrect to say

$$\{\text{alsation, boxer}\} = \{\text{breeds of dogs}\}$$

since there are many more breeds than the two that are given.

9

4. Describe a set that includes the given members of the following sets and state another member of each.

a) Hungary, Poland, Czechoslovakia, Bulgaria

b) 10, 20, 30, 40, 50

5. List the members in the given sets.

a) {prime numbers less than 12}

b) {letters used in the word ALGEBRA}

THE SYMBOLS ∈ AND ∉

The symbol ∈ means 'is a member of'
so that 'History is a member of the set of school subjects' may be written

$$\text{History} \in \{\text{school subjects}\}$$

Similarly the symbol ∉ means 'is not a member of'.
'Elm is not a breed of dog' may be written

$$\text{Elm} \notin \{\text{breeds of dogs}\}$$

EXERCISE 2b

Write each of the following statements in set notation.

1. John is a member of the set of boys' names.

2. English is a member of the set of school subjects.

3. June is not a day of the week.

4. A Jaguar is a British car.

5. Monday is not a member of the set of domestic furniture.

6. Curtain is not a member of the set of crockery.

State whether the following statements are true or false.

7. $32 \in \{\text{odd numbers}\}$

8. Rover $\notin \{\text{British cars}\}$

9. Washington $\in \{\text{American states}\}$

10. Washington $\in \{\text{capital cities}\}$

11. $1 \notin \{\text{prime numbers}\}$

FINITE AND INFINITE SETS

When we can write down, or count, all the members of a set, the set is called a *finite set,* e.g. A = {days of the week} is a finite set since there are seven days in a week. If we denote the number of members in the set A by $n(A)$, then $n(A) = 7$

Similarly if $B = \{5, 10, 15, 20, 25, 30\}$, $n(B) = 6$
and if $C = \{$letters in the alphabet$\}$, $n(C) = 26$

If there is no limit to the number of members in a set, the set is called an *infinite set* e.g. {even numbers} is an infinite set since we can go on adding 2 time and time again.

EXERCISE 2c

Are the following sets finite or infinite sets?

1. {odd numbers}

2. {the number of leaves on a particular tree}

3. {trees more than 60 m tall}

4. {the decimal numbers between 0 and 1}

5. {consonants}

6. {multiples of 5}

7. {men who have landed on the moon}

Find the number of elements in each of the following sets.

8. $A = \{$vowels$\}$

9. $B = \{$capital cities in Great Britain$\}$

10. $C = \{$prime numbers less than 20$\}$

Find $n(A)$ for each of the following sets.

11. $A = \{5, 10, 15, 20, 25, 30\}$

12. $A = \{a, e, i, o, u\}$

13. $A = \{$the consonants$\}$

14. $A = \{$odd numbers more than 10 but less than 26$\}$

15. $A = \{$players in a soccer team$\}$

16. $A = \{$players in a hockey team$\}$

EQUAL AND EMPTY SETS

Two sets are *equal* if they contain exactly the same elements, not necessarily in the same order.

e.g. if $A = \{$prime numbers greater than 2 but less than 9$\}$

and $B = \{$odd numbers between 2 and 8$\}$

then $A = B$ i.e. they are equal sets.

A set that has no members is called an *empty* or *null* set. It is denoted by ϕ or $\{\ \}$

EXERCISE 2d

State whether or not the following sets are equal.

1. $A = \{8, 4, 2, 12\}$, $B = \{2, 4, 6, 8\}$

2. $C = \{$letters of the alphabet except consonants$\}$
$D = \{$i, o, u, a, e$\}$

3. $X = \{$integers between 2 and 14 that are exactly divisible by 3 or 4$\{$
$Y = \{3, 4, 6, 8, 9, 12\}$

4. $A = \{$prime numbers less than 11$\}$
$B = \{1, 2, 3, 5, 7\}$

5. $C = \{$different letters in the word MANAGER$\}$
$D = \{$different letters in the word GERMAN$\}$

Determine whether or not the following sets are null sets.

6. $\{$animals that have travelled in space$\}$

7. $\{$multiples of 11 between 12 and 20$\}$

8. $\{$prime numbers less than 2$\}$

9. $\{$consonants$\}$

10. $\{$trees without leaves$\}$

UNIVERSAL SETS

Think of the set $\{$pupils in my class$\}$.

With this group of pupils in mind we might well think of several other sets,

i.e. $A = \{$pupils wearing spectacles$\}$

$B = \{$pupils wearing brown shoes$\}$

$C = \{$pupils with blue eyes$\}$

$D = \{$pupils more than 150 cm tall$\}$

We call the set {pupils in my class} a *universal set* for the sets *A*, *B*, *C* and *D*

All the members of *A*, *B*, *C* and *D* must be found in a universal set, but a universal set may contain other members as well.

We denote a universal set by $\mathscr{E}$ or U.

{pupils in my year at school} or {pupils in my school} would also be suitable universal sets for the sets *A*, *B*, *C* and *D* given above.

EXERCISE 2e

Suggest a universal set for {5, 10, 15, 20} and {6, 18, 24}

$\mathscr{E} = \{\text{integers}\}$

In questions 1 to 3 suggest a universal set for

1. {knife, dessert spoon}, {fork, spoon}

2. {10, 20, 30, 40}, {15, 25, 35}

3. {8, 12, 16, 20, 24}, {9, 12, 15, 18, 21, 24}

4. $\mathscr{E} = \{\text{integers from 1 to 20 inclusive}\}$
$A = \{\text{prime numbers}\}$
$B = \{\text{multiples of 3}\}$
List the sets *A* and *B*.

5. $\mathscr{E} = \{\text{letters of the alphabet}\}$
$A = \{\text{vowels}\}$
$B = \{\text{letters used in the word HYMNS}\}$
List the sets *A* and *B*.

6. $\mathscr{E} = \{\text{integers from 10 to 30 inclusive}\}$
$A = \{\text{prime numbers}\}$
$B = \{\text{multiples of 3}\}$
$C = \{\text{multiples of 4}\}$
List the sets *A*, *B* and *C*.

7. $\mathscr{E} = \{$positive integers less than 16$\}$
$A = \{$factors of 12$\}$
$B = \{$prime numbers$\}$
$C = \{$integers that are exactly divisible by 2 and by 3$\}$

List the sets A, B and C.

8. $\mathscr{E} = \{$letters in the word FINANCIAL$\}$
$A = \{$letters in the word CLINICAL$\}$
$B = \{$letters in the word FLAN$\}$

List the sets $\mathscr{E}, A$ and B.

9. $\mathscr{E} = \{x,$ a whole number, such that $4 \leqslant x \leqslant 20\}$
$A = \{$multiples of 5$\}$
$B = \{$multiples of 7$\}$
$C = \{$multiples of 4$\}$

List the sets A, B and C.

SUBSETS

If all the members of a set B are also members of a set A, then the set B is called a *subset* of the set A. This is written $B \subseteq A$. We use the symbol $\subseteq$ rather than $\subset$ if we don't know whether B could be equal to A.

Subsets that do not contain all the members of A are called *proper subsets*. If B is such a subset we write $B \subset A$.

EXERCISE 2f

If $A = \{$David, Edward, Fritz, Harold$\}$, write down all the subsets of A with exactly three members.

The subsets of A with exactly three members are
$\{$David, Edward, Fritz$\}$
$\{$David, Edward, Harold$\}$
$\{$David, Fritz, Harold$\}$
$\{$Edward, Fritz, Harold$\}$

1. If $A = \{$John, Jill, Peter, Audrey, Janet$\}$, write down all the subsets of A with exactly two female members.

2. If $N = \{$positive integers from 1 to 15 inclusive$\}$, list the following subsets of N.

$A = \{$even numbers from 1 to 15 inclusive$\}$
$B = \{$prime numbers less than 15$\}$
$C = \{$multiples of 3 that are less than or equal to 15$\}$

Do sets A and B have any element in common?

3. Give a subset with at least three members for each of the following sets.

a) $\{$European cities$\}$
b) $\{$British rivers$\}$
c) $\{$Commonwealth countries$\}$

4. If $A = \{$even numbers from 2 to 20 inclusive$\}$, list the following subsets of A.

$B = \{$multiples of 3$\}$
$C = \{$prime numbers$\}$
$D = \{$numbers greater than 12$\}$

VENN DIAGRAMS

In the Venn diagram the universal set ($\mathscr{E}$) is usually represented by a rectangle and the subsets of the universal set by circles within the rectangle.

If $\mathscr{E} = \{$families$\}$, $A = \{$families with one car$\}$
and $B = \{$families with more than one car$\}$

the Venn diagram would be

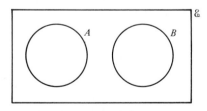

No family can have just one car and, at the same time, more than one car, i.e. A and B have no members in common.

Two such sets are called *disjoint sets*.

EXERCISE 2g

1.

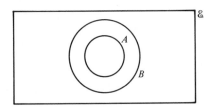

You are given the following information.
$\mathscr{E} = \{$pupils in my year$\}$
$A = \{$pupils in my class who are my friends$\}$
$B = \{$pupils in my class$\}$

a) Shade the region that shows the pupils in my class that are not my friends

b) Are all my friends in my class?

For each of questions 2 to 5 draw the diagram given below.

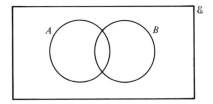

Using the following information shade the area asked for.
$\mathscr{E} = \{$pupils in my class$\}$
$A = \{$pupils in my class with fair hair$\}$
$B = \{$pupils in my class who are good at geography$\}$

2. $\{$pupils in my class with fair hair who are good at geography$\}$

3. $\{$pupils in my class who do not have fair hair$\}$

4. $\{$pupils in my class who do not have fair hair and who are not good at geography$\}$

5. $\{$pupils in my class with fair hair who are not good at geography$\}$

In questions 6 to 11

$\mathcal{E}$ = {pupils who attend my school}
A = {pupils who like coming to my school}
B = {pupils who are my friends}

In each case describe, in words, the shaded area.

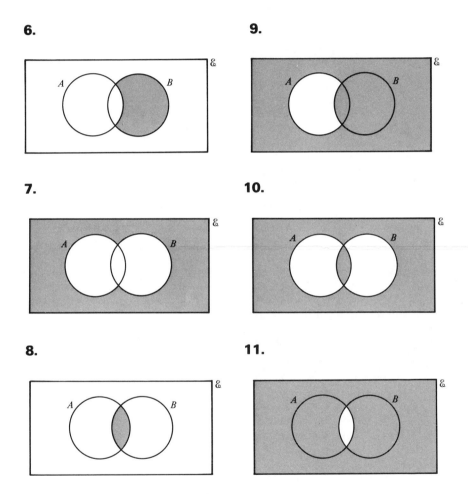

6.

9.

7.

10.

8.

11.

UNION OF TWO SETS

If we write down the set of all the members that are in either set A or set B we have what we call the *union* of the sets A and B.

The union of A and B is written $A \cup B$

EXERCISE 2h

$\mathscr{E} = \{1, 2, 3, 4, 5, 6, 7, 8\}$

If $A = \{2, 4, 6, 8\}$ and $B = \{1, 2, 3, 4, 5\}$ find $A \cup B$ illustrating these sets on a Venn diagram.

$A \cup B = \{1, 2, 3, 4, 5, 6, 8\}$

We could show this on a Venn diagram as follows.

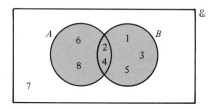

The shaded area represents the set $A \cup B$

In questions 1 to 5 find the union of the two given sets illustrating your answer on a Venn diagram.

1. $\mathscr{E} = \{\text{girls' names beginning with the letter J}\}$
$A = \{\text{Janet, Jill, Jenny}\}$ $B = \{\text{Judith, Janet, Jacky}\}$

2. $\mathscr{E} = \{\text{positive integers from 1 to 16 inclusive}\}$
$X = \{4, 8, 12, 16\}$ $Y = \{2, 6, 10, 14, 16\}$

3. $\mathscr{E} = \{\text{letters of the alphabet}\}$
$P = \{\text{letters in the word GEOMETRY}\}$
$Q = \{\text{letters in the word TRIGONOMETRY}\}$

4. $\mathscr{E} = \{\text{positive integers from 3 to 17 inclusive}\}$
$A = \{3, 5, 8, 12, 17\}$ $B = \{5, 7, 9, 11, 13, 15, 17\}$

5. $\mathscr{E} = \{\text{consonants}\}$
$P = \{p, q, r, s\}$ $Q = \{w, x, y, z\}$

6. Draw suitable Venn diagrams to show the unions of the following sets, and describe these unions in words as simply as possible.
a) $\mathscr{E} = \{\text{quadrilaterals}\}$ $A = \{\text{parallelograms}\}$ $B = \{\text{trapeziums}\}$
b) $\mathscr{E} = \{\text{angles}\}$ $P = \{\text{obtuse angles}\}$ $Q = \{\text{reflex angles}\}$

INTERSECTION OF TWO SETS

The set of all the members that are members both of set A and of set B is called the intersection of A and B, and is written $A \cap B$

EXERCISE 2i

$\mathscr{E} = \{\text{integers from 1 to 12 inclusive}\}$
If $A = \{1, 2, 3, 4, 5, 6, 7, 8\}$ and $B = \{1, 2, 3, 5, 7, 11\}$ find $A \cap B$
illustrating it on a Venn diagram.

$A \cap B = \{1, 2, 3, 5, 7\}$

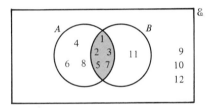

The shaded area represents the set $A \cap B$

Draw suitable Venn diagrams to show the intersections of the following sets. In each case write down the intersection in set notation.

1. $\mathscr{E} = \{\text{integers from 4 to 12 inclusive}\}$
$X = \{4, 5, 6, 7, 10\}$ $Y = \{5, 7, 11\}$

2. $\mathscr{E} = \text{colours of the rainbow}\}$
$A = \{\text{red, orange, yellow}\}$ $B = \{\text{blue, red, violet}\}$

3. $\mathscr{E} = \{\text{positive whole numbers}\}$
$C = \{\text{positive whole numbers that divide exactly into 24}\}$
$D = \{\text{positive whole numbers that divide exactly into 28}\}$

4. $\mathscr{E} = \{\text{letters of the alphabet}\}$
$P = \{\text{different letters in the word BRITISH}\}$
$Q = \{\text{different letters in the word IRISH}\}$

5. $\mathcal{E} = \{\text{positive whole numbers}\}$
$X = \{\text{positive whole numbers less than } 10\}$
$Y = \{\text{positive odd numbers less than } 12\}$

6. $\mathcal{E} = \{\text{integers less than } 25\}$
$A = \{\text{multiples of 3 between 7 and 23}\}$
$B = \{\text{multiples of 4 between 7 and 23}\}$

SIMPLE PROBLEMS INVOLVING VENN DIAGRAMS

EXERCISE 2j

If $\mathcal{E} = \{\text{girls in my class}\}$
$A = \{\text{girls who play hockey}\} = \{\text{Helen, Bina, Moira, Sara, Lana}\}$
and $B = \{\text{girls who play tennis}\} = \{\text{Kath, Sara, Helen, Maria}\}$
illustrate A and B on a Venn diagram. Use this diagram to write down the following sets:
a) $\{\text{girls who play both hockey and tennis}\}$
b) $\{\text{girls who play hockey but not tennis}\}$
c) $\{\text{girls who play tennis but not hockey}\}$

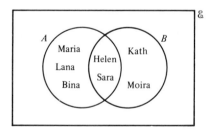

From the Venn diagram
a) $\{\text{girls who play both hockey and tennis}\} = \{\text{Helen, Sara}\}$
b) $\{\text{girls who play hockey but not tennis}\} = \{\text{Moira, Lana, Bina}\}$
c) $\{\text{girls who play tennis but not hockey}\} = \{\text{Kath, Maria}\}$

1. $\mathscr{E} = \{$the pupils in a class$\}$
$X = \{$pupils who like history$\}$
$Y = \{$pupils who like geography$\}$

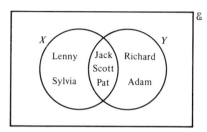

List the set of pupils who
a) like history but not geography   b) like geography but not history
c) like both subjects.

2. $\mathscr{E} = \{$boys in my class$\}$
$A = \{$boys who play soccer$\}$ $B = \{$boys who play rugby$\}$

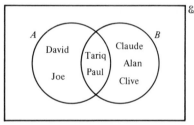

Write down the sets of boys who
a) play soccer   b) play both games   c) play rugby but not soccer.

3. $\mathscr{E} = \{$my friends$\}$
$P = \{$friends who wear glasses$\}$ $Q = \{$friends who wear brown shoes$\}$

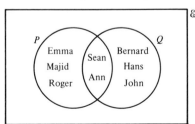

List all my friends who
a) wear glasses   b) wear glasses but not brown shoes
c) wear both glasses and brown shoes.

4. $\mathscr{E} = \{\text{whole numbers from 1 to 14 inclusive}\}$
$A = \{\text{even numbers between 3 and 13}\}$
$B = \{\text{multiples of 3 between 1 and 14}\}$

Illustrate this information on a Venn diagram and hence write down

a) the even numbers between 3 and 13 that are multiples of 3

b) multiples of 3 between 1 and 14 that are odd.

5. $P = \{\text{different letters in the word SCHOOL}\}$
$Q = \{\text{different letters in the word SQUASH}\}$

Show these on a Venn diagram and hence write down

a) the letters that are in both words

b) the letters found only in one word

c) the letters used in the word SCHOOL but not in SQUASH.

$\mathscr{E} = \{\text{months of the year}\}$
$A = \{\text{months of the year beginning with the letter J}\}$
$B = \{\text{months of the year ending with the letter Y}\}$
Show these sets on a Venn diagram.
Hence find   a) $A \cup B$ b) $A \cap B$, describing each of the sets.

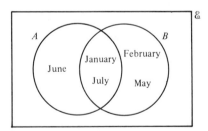

$A \cup B = \{\text{January, February, May, June, July}\}$

$= \{\text{months of the year } \textit{either} \text{ beginning with J } \textit{or} \text{ ending with Y}\}$

$A \cap B = \{\text{January, July}\}$

$= \{\text{months of the year } \textit{both} \text{ beginning with J } \textit{and} \text{ ending with Y}\}$

6. $\mathcal{E} = \{$letters of the alphabet$\}$
     $P = \{$letters used in the word LIBERAL$\}$
     $Q = \{$letters used in the word LABOUR$\}$

Show these on a Venn diagram.
Hence find   a) $P \cap Q$ b) $P \cup Q$ describing each of these sets.

7. $\mathcal{E} = \{$whole numbers less than 12$\}$
     $C = \{$prime numbers$\}$ $D = \{$odd numbers$\}$

Hence find   a) $C \cap D$ b) $C \cup D$

8. $\mathcal{E} = \{$whole numbers from 1 to 35 inclusive$\}$
     $R = \{$multiples of 4$\}$ $S = \{$multiples of 6$\}$

Show these on a Venn diagram.
Hence find   a) $R \cup S$ b) $R \cap S$

COMPLEMENT

If $\mathcal{E} = \{$pupils in my school$\}$
and $A = \{$pupils who are good at games$\}$
then the *complement of A* is the set of all the members of $\mathcal{E}$ that are *not* members of A.

In this case, the complement of A is
$\{$pupils in my school who are not good at games$\}$

The complement of A is denoted by A'.

Similarly if $\mathcal{E} = \{$the whole numbers from 1 to 10 inclusive$\}$
       and $A = \{$odd numbers$\} = \{1, 3, 5, 7, 9\}$
the complement of A, i.e. A', is $\{2, 4, 6, 8, 10\} = \{$even numbers$\}$

EXERCISE 2k

> Give the complement of P where
> $P = \{$Thursday, Friday$\}$ if $\mathcal{E} = \{$days of the week$\}$
>
> $P' = \{$Monday, Tuesday, Wednesday, Saturday, Sunday$\}$

Give the complement of each of the following sets.

1. $A = \{5, 15, 25\}$ if $\mathcal{E} = \{5, 10, 15, 20, 25\}$

2. $B = \{7, 8, 9, 10\}$ if $\mathcal{E} = \{5, 6, 7, 8, 9, 10, 11\}$

3. $V = \{a, e, i, o, u\}$ if $\mathcal{E} = \{\text{letters of the alphabet}\}$

4. $P = \{\text{the consonants}\}$ if $\mathcal{E} = \{\text{letters of the alphabet}\}$

5. $A = \{\text{Monday, Wednesday, Friday}\}$ if $\mathcal{E} = \{\text{days of the week}\}$

6. $X = \{\text{children}\}$ if $\mathcal{E} = \{\text{human beings}\}$

7. $M = \{\text{British motor cars}\}$ if $\mathcal{E} = \{\text{motor cars}\}$

8. $S = \{\text{male tennis players}\}$ if $\mathcal{E} = \{\text{tennis players}\}$

9. $C = \{\text{British capital cities}\}$ if $\mathcal{E} = \{\text{capital cities of the United Kingdom}\}$

10. $D = \{\text{squares}\}$ if $\mathcal{E} = \{\text{quadrilaterals}\}$

11. $E = \{\text{adults over 80 years old}\}$ if $\mathcal{E} = \{\text{adults}\}$

12. $F = \{\text{male doctors}\}$ if $\mathcal{E} = \{\text{female doctors}\}$

If $A = \{\text{men}\}$ and $A' = \{\text{women}\}$ what is $\mathcal{E}$?

$$\mathcal{E} = \{\text{adults}\}$$

---

13. If $A = \{\text{homes with television sets}\}$ and $A' = \{\text{homes without television sets}\}$ what is $\mathcal{E}$?

14. If $A = \{\text{vowels}\}$ and $A' = \{\text{consonants}\}$ what is $\mathcal{E}$?

15. If $X = \{a, b, c, d, e\}$ and $X' = \{f, g, h, i, j\}$ what is $\mathcal{E}$?

16.

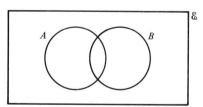

a) Copy the Venn diagram and shade the region representing A'.

b) Copy the Venn diagram and shade the region representing B'.

HARDER PROBLEMS

EXERCISE 2I

Given $\mathscr{E} = \{1, 2, 3, 4, 5, 6, 7, 8, 9, 10\}$
$A = \{\text{odd numbers}\} = \{1, 3, 5, 7, 9\}$
$B = \{\text{multiples of } 3\} = \{3, 6, 9\}$
Show these sets on a Venn diagram.
Use your diagram to list the following sets:
a) A' b) B' c) $A \cup B$ d) the complement of $A \cup B$,
i.e. $(A \cup B)'$ e) $A' \cap B'$

(Each of the numbers in the given sets is placed in the correct position on the following Venn diagram.)

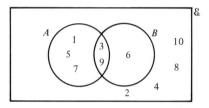

From the diagram a) $A' = \{2, 4, 6, 8, 10\}$
　　　　　　　　b) $B' = \{1, 2, 4, 5, 7, 8, 10\}$
　　　　　　　　c) $A \cup B = \{1, 3, 5, 6, 7, 9\}$
　　　　　　　　d) $(A \cup B)' = \{2, 4, 8, 10\}$
　　　　　　　　e) $A' \cap B' = \{2, 4, 8, 10\}$

1. $\mathscr{E} = \{1, 2, 3, 4, 5\}$ $A = \{1, 2, 3, 5\}$ $B = \{2, 4\}$
Show the sets on a Venn diagram and use it to find
a) A' b) B' c) $A \cup B$ d) $(A \cup B)'$ e) $A' \cap B'$

2. $\mathscr{E} = \{\text{whole numbers less than } 17\}$ $P = \{\text{multiples of } 3\}$
$Q = \{\text{even numbers}\}$
Show these on a Venn diagram and use it to find
a) P' b) Q' c) $P \cup Q$ d) $(P \cup Q)'$ e) $P' \cap Q'$

3. $\mathscr{E} = \{1, 2, 3, 4, 5, 6, 7, 8, 9, 10, 11, 12\}$
$A = \{\text{multiples of } 4\}$ $B = \{\text{even numbers}\}$
Show these sets on a Venn diagram and use this diagram to list the sets
a) A' b) B' c) $A \cup B$ d) $(A \cup B)'$ e) $A' \cap B'$

4. $\mathscr{E} = \{$whole numbers from 10 to 25$\}$

$P = \{$multiples of 4$\}$ $Q = \{$multiples of 5$\}$

Show these sets on a Venn diagram and use this diagram to list the sets

a) P' b) Q' c) $P \cup Q$ d) $(P \cup Q)'$ e) $P' \cap Q'$

5. $\mathscr{E} = \{$different letters in the word GENERAL$\}$

$A = \{$different letters in the word ANGEL$\}$

$B = \{$different letters in the word LEAN$\}$

Show these sets on a Venn diagram and use this diagram to list the sets

a) A' b) B' c) $A \cap B$ d) $A \cup B$ e) $(A \cap B)'$

f) $A' \cap B'$

6. $\mathscr{E} = \{$p, q, r, s, t, u, v, w$\}$

$X = \{$r, s, t, w$\}$ $Y = \{$q, s, t, u, v$\}$

Show $\mathscr{E}$, X and Y on a Venn diagram entering all the members. Hence list the sets

a) X' b) Y' c) $X' \cap Y'$ d) $X \cup Y$ e) $(X \cup Y)'$

Which two sets are equal?

7. $\mathscr{E} = \{1, 2, 3, 4, 5, 6, 7, 8, 9, 10, 11, 12\}$

$X = \{$factors of 12$\}$ $Y = \{$even numbers$\}$

Show $\mathscr{E}$, X and Y on a Venn diagram entering all the members. Hence list the sets

a) X' b) Y' c) $X' \cap Y'$

d) $X' \cup Y'$ e) $X \cup Y$ f) $(X \cup Y)'$

Which two sets are equal ?

8. Draw the diagram given opposite six times. Use shading to illustrate each of the following sets.

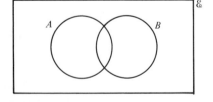

a) A' b) B' c) $A' \cap B'$

d) $A' \cup B'$ e) $A \cup B$

f) $(A \cup B)'$

9. $\mathscr{E} = \{$different letters in the word MATHEMATICS$\}$

$A = \{$different letters in the word ATTIC$\}$

$B = \{$different letters in the word TASTE$\}$

Show $\mathscr{E}$, A and B on a Venn diagram entering all the elements. Hence list the sets

a) A' b) B' c) $A \cup B$

d) $(A \cup B)'$ e) $A' \cup B'$ f) $A' \cap B'$

10. $\mathscr{E} = \{$pupils in my class$\}$
$P = \{$those with compasses$\}$
$Q = \{$those with protractors$\}$
Describe   a) P' b) Q' c) $P' \cap Q'$ d) $(P \cup Q)'$ e) $P \cup Q$

NUMBER OF MEMBERS

EXERCISE 2m

Illustrate on a Venn diagram the sets A and B if
$A = \{$Sunday, Monday, Tuesday, Wednesday$\}$
$B = \{$Wednesday, Thursday, Friday, Saturday, Sunday$\}$
Use your diagram to find
a) $n(A)$ b) $n(B)$ c) $n(A \cup B)$ d) $n(A \cap B)$

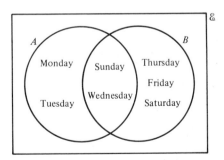

Counting the number of elements in the various regions gives
a) $n(A) = 4$ b) $n(B) = 5$
c) $n(A \cup B) = 7$ d) $n(A \cap B) = 2$

In questions 1 to 6 count the number of elements in the various regions to find

a) $n(A)$ b) $n(B)$ c) $n(A \cup B)$ $n(A \cap B)$

1.

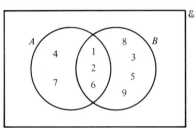

4.

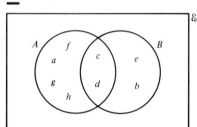

2.

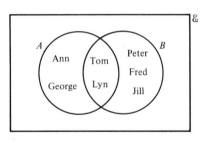

5.

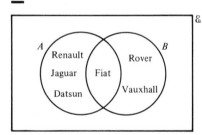

3.

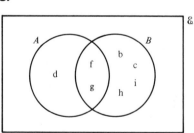

6.

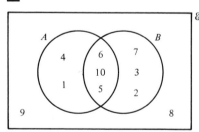

In the remaining questions in this exercise, the numbers in the various regions of the Venn diagrams show the *number of elements in,* or *members of, the set in that region.*

In questions 7 to 10 use the information given in the Venn diagrams to find

a) $n(X)$ b) $n(Y)$ c) $n(X \cup Y)$ d) $n(X \cap Y)$

7.

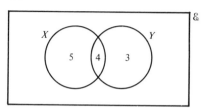

9.

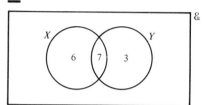

8.

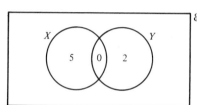

10.

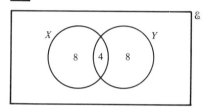

Use the information given in the Venn diagram to find
$n(A)$, $n(B)$, $n(A')$, $n(B')$, $n(A \cup B)$, $n(A \cap B)$, $n(A' \cup B')$,
$n[(A \cap B)']$ for the given sets.

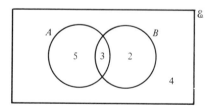

The numbers in the regions show the number of elements in that region.

$n(A) = 8$ $n(B) = 5$
$n(A') = 6$ $n(B') = 9$
$n(A \cup B) = 10$ $n(A \cap B) = 3$
$n(A' \cup B') = 11$ $n[(A \cap B)'] = 11$

Use the information given in the following Venn diagrams to find $n(A)$, $n(B)$, $n(A')$, $n(B')$, $n(A \cup B)$, $n(A \cap B)$, $n(A' \cup B')$ and $n[(A \cap B)']$ for each of the given pairs of sets.

11.

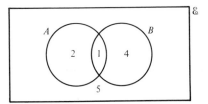

14.

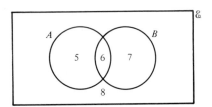

12.

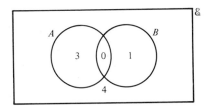

15.

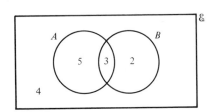

13.

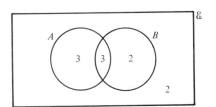

16.

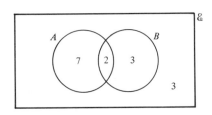

PROBLEMS

EXERCISE 2n

The Venn diagram shows how many pupils in a class have telephones (T) and video recorders (V)

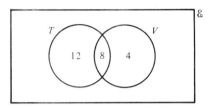

How many pupils have
a) both a telephone and a video recorder
b) a telephone
c) a telephone and/or a video recorder
d) a video recorder but not a telephone?

a) The number of pupils with both is 8.
b) The number of pupils with a telephone is $12 + 8$, i.e. 20.
c) The number of pupils with at least one of the two
is $12 + 8 + 4$, i.e. 24.
d) 4 pupils have a video recorder but not a telephone.

1. The Venn diagram shows the numbers of boys in a class who play soccer (S) and who play cricket (C).

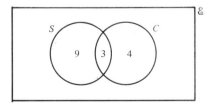

How many boys play   a) both games   b) only cricket   c) soccer
d) exactly one of these games?

2. The students in a form were asked if they did any cooking (C) or dressmaking (D) at home. Their replies are shown in the Venn diagram.

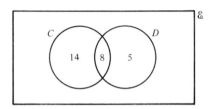

If all the students in the form took part in at least one of these activities, how many students

a) are there in the form   b) did only cooking

c) did both   d) did exactly one of these activities?

3. In a group of 24 children, each had a dog or a cat or both. If 18 kept a dog and 5 of these also kept a cat, show this information on a Venn diagram and hence find the number of children who kept

a) a cat   b) only a dog   c) just one of these as a pet.

4. A group of 50 television addicts were asked if they watched BBC and ITV. The replies revealed that 21 watched both channels but 9 watched only ITV. Show this information on a Venn diagram and use it to find the numbers of viewers who

a) watched BBC   b) did not watch ITV   c) watched only one channel.

5. In a Youth Club 35 teenagers said that they went to football matches, discos or both. Of the 22 who said they went to football matches, 12 said they also went to discos. Show this information on a Venn diagram. How many went to football matches or discos, but not to both?

6. There are 28 pupils in a form, all of whom take history or geography or both. If 14 take history, 5 of whom also take geography, show this information on a Venn diagram and hence find the number of pupils who take

a) geography   b) history but not geography

c) just one of these subjects.

7. In a squad of 35 rugby players 20 said that they would play in the backs and 8 said that they would play in either the backs or the forwards. Show this information on a Venn diagram. How many more were willing to play in the forwards than in the backs?

The Venn diagram shows how many houses in a street have double glazing (D) and how many have new front doors (N).

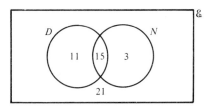

How many houses
a) are there in the street
b) do not have a new front door
c) have either a new front door or double glazing but not both ?

a) Number of houses in the street is $11 + 15 + 3 + 21$, i.e. 50

b) Number of houses without a new front door is $11 + 21$, i.e. 32

c) The numbers with either a new front door or double glazing but not both is $11 + 3$, i.e. 14

8. The Venn diagram shows how many pupils in a class passed the English examination (E) and how many passed the mathematics examination (M).

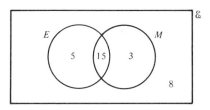

How many pupils

a) passed in only one examination

b) did not pass in English

c) passed in at least one examination ?

9. The Venn diagram shows how many pupils in a class kept goldfish (G), budgerigars (B) or both.

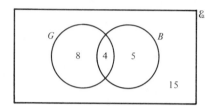

How many pupils
a) were there in the class   b) did not have a budgerigar
c) had at least one of these pets?

10. The passengers on a coach were questioned about the newspapers they bought each day.

  3 bought both a morning and evening newspaper.

  15 bought a morning newspaper.

  8 bought an evening newspaper.

  8 did not buy a newspaper at all.

Show this information on a Venn diagram.

a) How many passengers were there on the coach?

b) How many passengers bought a newspaper at some time in the day?

11. One evening all 78 members of a Youth Club were asked whether they liked swimming (S) and/or dancing (D). It was found that 34 liked swimming, 41 liked dancing and 19 liked both. Show this information on a Venn diagram. How many were

a) swimmers but not dancers   b) dancers or swimmers but not both
c) neither dancers nor swimmers?

12. During April, 36 cars were taken to a Testing Station for an MOT certificate. The results showed that 8 had defective brakes and lights, 10 had defective brakes, and 13 had defective lights. How many cars

a) failed the test   b) passed the test   c) had exactly one defect?

13. The 32 pupils in a class were asked whether they studied French or art or both. It was found that 8 studied both, 13 studied French and 6 did not study either subject. How many pupils studied

a) art but not French   b) French or art but not both?

The following exercise extends the ideas that we have used so far, to three intersecting sets.

EXERCISE 2p

In a certain group of pupils, some are in one or more of the school swimming, debating and trampoline teams. The Venn diagram shows these numbers where

$S = \{$those in the swimming teams$\}$
$D = \{$those in debating teams$\}$
and $T = \{$those in trampoline teams$\}$

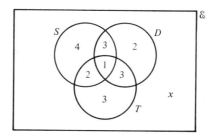

How many take part in

a) debating teams only
b) at least one type of team
c) exactly two types of teams?

If there are 24 in the group find

d) the value of x
e) the number who do not belong to a debating team

   a) 2 take part in debating teams and nothing else.
   b) The number taking part in at least one type of team is
      $4 + 2 + 3 + 3 + 2 + 3 + 1$ i.e. 18.
   c) Taking part in exactly two types of teams are $3 + 3 + 2$
      i.e. 8 pupils.
   d) Since $18 + x = 24$, 6 pupils do not belong to any team.
   e) The number who do not belong to any debating team
      is $24 - (3 + 2 + 3 + 1)$ i.e. 15 pupils.

1.

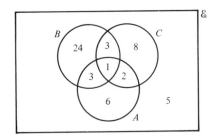

The Venn diagram shows the methods of transport used by the staff of a London store where
$A = \{$those using a taxi$\}$, $B = \{$those using a bus$\}$
and $C = \{$those using a car$\}$.

a) How many members of staff does the store employ?

b) How many of the staff
    i) use exactly one means of transport
    ii) use more than one means of transport
    iii) do not use a bus?

2. In a certain group of students some are in one or more of the school hockey, netball or swimming teams. The Venn diagram shows these numbers where
$H = \{$those in the hockey team$\}$, $N = \{$those in the netball team$\}$ and $S = \{$those in the swimming team$\}$.

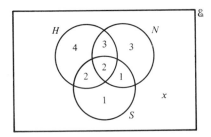

How many pupils are members of

a) the hockey team only

b) at least two teams

c) exactly one team

d) the netball team.

If there are 21 pupils in the group, what is the value of x?

Describe the sets   i) $H \cap N$ ii) $N \cap S'$ iii) N'

3. Copy the following Venn diagram.

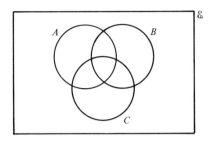

If $\mathscr{E} = \{1, 2, 3, 4, 5, 6, 7, 8, 9, 10, 11, 12\}$
   $A = \{\text{even numbers}\}$
   $B = \{\text{multiples of 4}\}$
and $C = \{\text{multiples of 3}\}$

place each of the numbers in its correct region. Describe in set language any empty sets.

4. The universal set is $\mathscr{E}$ where $\mathscr{E} = \{\text{positive integers from 1 to 24 inclusive}\}$. X, Y and Z are subsets of $\mathscr{E}$ and are defined as follows: $X = \{\text{multiples of 4}\}$, $Y = \{\text{multiples of 5}\}$, and $Z = \{\text{prime numbers}\}$. Draw a Venn diagram to illustrate this information and use it to list the elements of the sets

a) $X \cup Y$ b) $X' \cap Z$ c) $X' \cap Y' \cap Z'$ d) $(X \cup Y) \cap (Y \cup Z)$

5. Given $\mathscr{E} = \{\text{letters of the alphabet}\}$, $A = \{\text{letters used to make the word ALGEBRA}\}$, $B = \{\text{letters used to make the word ARITHMETIC}\}$, $C = \{\text{letters used to make the word GEOMETRY}\}$. Write down

a) the elements in the set $(A \cup B) \cap C$ b) $n(B \cup C)$.

6. If $\mathscr{E} = \{\text{quadrilaterals}\}$, $A = \{\text{parallelograms}\}$, $B = \{\text{rectangles}\}$ and $C = \{\text{squares}\}$, draw a Venn diagram to illustrate the connection between the sets.

7. If $\mathscr{E} = \{\text{triangles}\}$, $X = \{\text{right-angled triangles}\}$, $Y = \{\text{equilateral triangles}\}$ and $Z = \{\text{isosceles triangles}\}$, draw a Venn diagram to show the relationship between the sets.

Describe the elements of   a) $X \cap Z$ b) $X' \cap Y$ c) $X \cup Y$

8. Draw a Venn diagram to show three sets A, B and C in a universal set $\mathcal{E}$. Enter numbers in the correct parts of your diagram using the following information.

$n(A \cap B \cap C) = 2$, $n(A \cap B) = 7$, $n(B \cap C) = 6$, $n(A \cap C) = 8$, $n(A) = 16$, $n(B) = 20$, $n(C) = 19$, and $n(\mathcal{E}) = 50$.

Use your diagram to find

a) $n(A' \cap C')$ b) $n(A \cup B')$ c) $n(A' \cap B' \cap C')$

9.

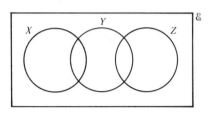

The Venn diagram shows three sets X, Y and Z contained within a universal set $\mathcal{E}$. Enter numbers in the correct regions of your diagram using the following information.

$n(X \cap Y) = 3$, $n(Y \cap Z) = 4$, $n(X) = 8$, $n(Y) = 18$, $n(Z) = 10$, and $n(\mathcal{E}) = 35$.

Use your diagram to find

a) $n(X \cap Y \cap Z)$ b) $n(X' \cup Y)$ c) $n(X' \cap Z')$

10. Forty travellers were questioned about the various methods of transport they had used the previous day. Every one of them had used at least one of the methods shown in the Venn diagram.

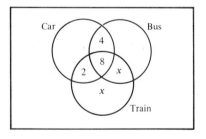

Of those questioned 8 had used all three methods of transport, 4 had travelled by bus and car only, and 2 by car and train only. The number (x) who had travelled by train only was equal to the number who had travelled by bus and train only.

If 20 travellers had used a train and 33 had used a bus find

a) the value of x

b) the number who travelled by bus only

c) the number who used exactly two methods of transport

d) the number who travelled by car only.

11. Some tests were held in the fourth form in physics (*P*), chemistry (*C*) and biology (*B*). Forty-five pupils took part and the recorded passes were as follows: physics 24, chemistry 25, biology 30, physics and chemistry but not biology 3, physics and biology but not chemistry 6, biology and chemistry but not physics 7. If 10 pupils passed in all three and 2 pupils failed in all three draw a Venn diagram to illustrate this information entering the correct number in each region.

Use your diagram to find the number of pupils who

a) passed in chemistry only

b) passed in more than one of these subjects

c) passed in exactly one of these subjects

d) did not pass in either biology or chemistry.

12. In a particular sixth form all the pupils belong to one or more of the groups studying arts subjects (*A*), science subjects (*B*) and craft subjects (*C*) as indicated in the Venn diagram below.

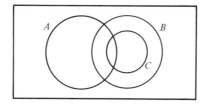

Which of the following statements are true ?

a) some students do not study an arts subject

b) some students study subjects of all three types

c) all students studying a craft subject also study an arts subject

d) all students studying an arts subject also study a craft subject

e) every student studying an arts subject also studies a science subject

f) every student studying a craft subject also studies a science subject.

3 PRISMS AND PYRAMIDS

VOLUME OF A CUBOID

The volume of a cuboid = length × breadth × height
The three measurements must be expressed in the same unit before multiplying.

EXERCISE 3a

Find the volume of a cuboid measuring 42 cm by 1.2 m by 382 mm. Give your answer in cubic metres.

$$42\,\text{cm} = 0.42\,\text{m}$$

$$382\,\text{mm} = 0.382\,\text{m}$$

$$\text{Volume} = 1.2 \times 0.42 \times 0.382\,\text{m}^3$$

$$= 0.1925\,\text{m}^3$$

$$= 0.193\,\text{m}^3 \quad \text{correct to 3 s.f.}$$

Find the volumes of the cuboids whose measurements are given in questions 1 to 5. Give your answers in the units indicated in the brackets.

1. 62 cm by 48 cm by 0.12 m   (cm³)

2. 1.3 cm by 62 mm by 1.7 cm   (cm³)

3. 420 cm by 500 cm by 620 cm   (m³)

4. 0.03 m by 0.16 m by 0.09 m   (cm³)

5. 1.6 cm by 1.5 cm by 7 mm   (mm³)

6. a) How many cubic centimetres are there in 1 m³ ?
   b) Express 4.23 m³ in cubic centimetres.

7. a) How many cubic millimetres are there in 1 cm³ ?
   b) Express 628 mm³ in cubic centimetres.

In questions 8 to 13, express the given volumes in the units indicated in brackets. Remember that 1 litre $= 1000 \text{ cm}^3$

8. 4200 cm^3 (litres)

9. 0.048 m^3 (cm³)

10. $75\,000\,000 \text{ cm}^3$ (m³)

11. $432\,000 \text{ cm}^3$ (m³)

12. 7800 mm^3 (cm³)

13. 42 cm^3 (mm³)

Find the length of a cuboid of volume 18 cm^3 whose breadth is 5.2 cm and whose height is 12 mm.

$$\text{Height} = 12 \text{ mm}$$
$$= 1.2 \text{ cm}$$
$$\text{Volume} = \text{length} \times \text{breadth} \times \text{height}$$
$$18 = l \times 5.2 \times 1.2$$
$$\frac{18}{5.2 \times 1.2} = l$$
$$\therefore \qquad l = 2.884$$

The length is 2.88 cm   correct to 3 s.f.

Find the missing measurements for each of the following cuboids.

	Volume	Length	Breadth	Height
14.	128 cm^3	4.8 cm	2 cm	cm
15.	24 m^3	m	0.7 m	15 m
16.	32 cm^3	cm	8 cm	64 mm
17.	241 mm^3	2.2 cm	mm	5.2 mm
18.	cm³	6.2 cm	0.3 m	190 mm

19. A block of metal measures 44 cm by 10 cm by 6 cm. It is melted down and formed into a cuboid which is 4 cm wide and 33 cm high. How long is the new block?

20. Rain falls on a flat roof measuring 3 m by 4 m and runs off into a rectangular tank which is 0.4 m long and 0.6 m wide. 1 cm of rain falls on the roof.

a) Find the volume of rain falling on the roof.

b) If the tank is empty to start with, how deep is the water in it when the rain stops ?

21.

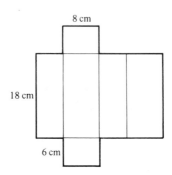

The diagram shows the net for a cuboid. Find the volume of the cuboid.

VOLUME OF A PRISM

A solid whose cross-section is the same all through is called a *prism.*

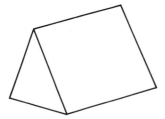
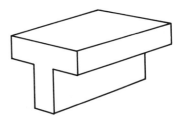

The volume of a solid whose cross-section is the same all the way through is given by

$$\text{volume} = \text{area of cross-section} \times \text{length}$$

EXERCISE 3b

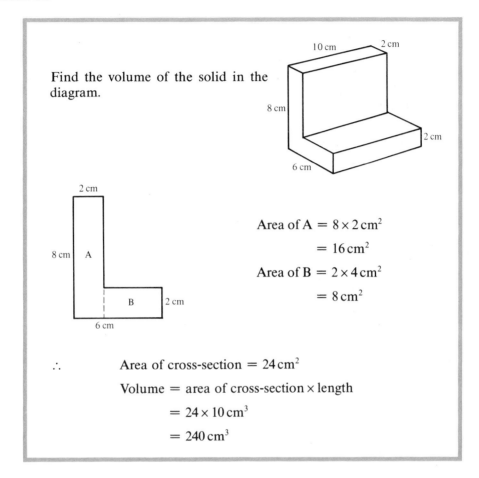

Find the volume of the solid in the diagram.

Area of A $= 8 \times 2 \, \text{cm}^2$

$= 16 \, \text{cm}^2$

Area of B $= 2 \times 4 \, \text{cm}^2$

$= 8 \, \text{cm}^2$

∴ Area of cross-section $= 24 \, \text{cm}^2$

Volume $=$ area of cross-section $\times$ length

$= 24 \times 10 \, \text{cm}^3$

$= 240 \, \text{cm}^3$

Find the volumes of the solids illustrated in questions 1 to 6. In each case draw a diagram of the cross-section but do not draw a picture of the solid.

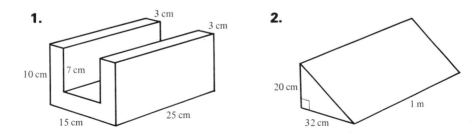

1.

2.

3.

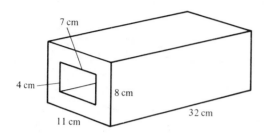

5.

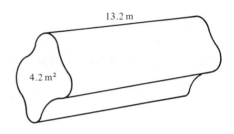

4.

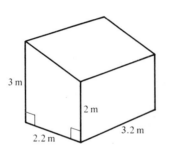

6.

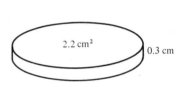

In each question from 7 to 10, find the volume of the solid whose cross-section and length are given. Give the answer in the unit indicated in brackets.

7.

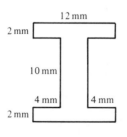

Length = 1 m
(cm³)

8.

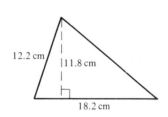

. Length = 24 cm
(cm³)

9.

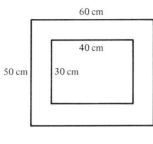

Length = 4 m
(m³)

10.

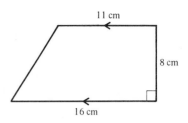

Length = 16 cm
(cm³)

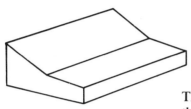

The volume of the solid shown in the diagram is 144 cm³ and the area of its cross-section is 14 cm². Find its length.

Let its length be l cm

$$\text{Volume} = \text{area of cross-section} \times \text{length}$$

$$144 = 14 \times l$$

$$l = \frac{144}{14}$$

$$= 10.28$$

i.e. the length is 10.3 cm   correct to 3 s.f.

11. The volume of a solid of uniform cross-section is 72 cm³. The area of its cross-section is 8 cm². Find the length of the solid.

12. The volume of a solid of uniform cross-section is 32 m³. Its length is 10 m. Find the area of its cross-section.

13.

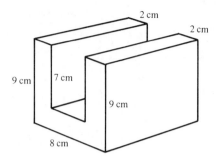

The volume of the solid is 396 cm³.

Find   a) the area of the cross-section

b) the length of the solid.

14.

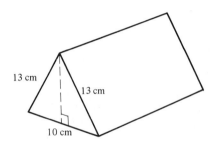

The cross-section of the solid is an isosceles triangle. The volume of the solid is 1200 cm³.

Find   a) the height of the triangle

b) the area of the cross-section

c) the length of the solid.

15.

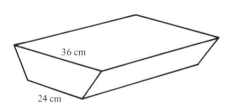

The cross-section of the solid is a trapezium. The height of the trapezium is 10 cm and the volume of the solid is 7800 cm³.

Find the length of the solid.

16. A drop of oil of volume 2.5 cm³ is dropped on to a flat surface and spreads out to form a pool of even thickness and area 50 cm².

How thick is the oil   a) in centimetres   b) in millimetres ?

17.

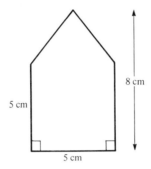

A cuboid of metal measuring 6 cm by 8.2 cm by 9.5 cm is recast into the shape of a prism. The cross-section of the prism is shown in the diagram. How long is the prism ?

Water comes out of a pipe of cross-section 3.2 cm² at a speed of 0.5 m/s. How much water is delivered by the pipe in one second?

(Imagine 0.5 m of pipe being emptied in 1 second)

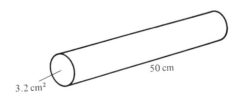

Volume = area of cross-section × length

= 3.2 × 50 cm³

= 160 cm³

∴ 160 cm³ of water is delivered in 1 second

18. The cross-section of a pipe is 4.8 cm². If water comes out of the pipe at 30 cm/s, how much water is delivered in 1 second?

19. Water comes out of a pipe at 60 cm/s. The cross-section of the pipe is a circle of radius 0.5 cm.

How much water is delivered   a) in 1 second   b) in 1 minute?

20.

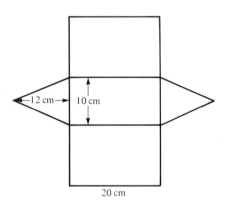

The diagram shows the net for a prism with an isosceles triangular cross-section.

Find   a) the lengths of the sides of the triangular cross-section.

b) the area of the triangular ends

c) the volume of the prism.

21.

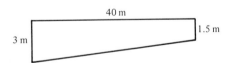

The diagram shows the side view of a swimming bath of width 25 m.

a) Find the volume of water in the bath when it is full. Give your answer in cubic metres.

b) The bath is emptied through a pipe whose cross-sectional area is 200 cm²; the water runs out at 1.5 m/s. What volume of water is removed in 1 second?

c) Find how long it would take to empty the bath if four similar pipes are used each removing water at the same steady rate as in (b).

VOLUME OF A PYRAMID

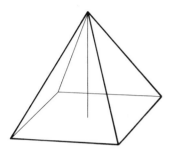 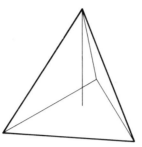 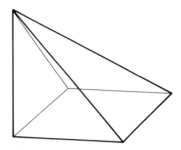

We can show that the volume of a pyramid is given by

$$\text{Volume} = \tfrac{1}{3} \text{ area of base} \times \text{perpendicular height}$$

EXERCISE 3c

Find the volume of the pyramid in the diagram. Its height is 7 cm and its base is a rectangle.

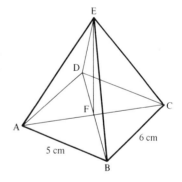

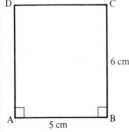

Area of base $= 5 \times 6 \text{ cm}$

$$= 30 \text{ cm}^2$$

Volume $= \tfrac{1}{3}$ area of base $\times$ perpendicular height

$$= \tfrac{1}{3} \times 30^{10} \times 7 \text{ cm}^3$$

$$= 70 \text{ cm}^3$$

Find the volumes of the pyramids in questions 1 to 7.

1.

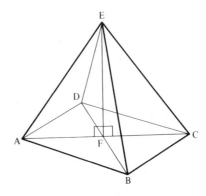

The base ABCD is a rectangle.
AB = 8 cm, BC = 4.5 cm and EF = 6 cm

2.

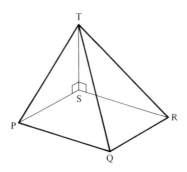

The base PQRS is a rectangle.
PQ = 20 cm, QR = 12 cm and TS = 8 cm

3.

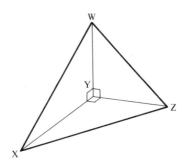

The base is a triangle XYZ.
$X\widehat{Y}Z = 90°$, XY = 10 cm, YZ = 8 cm and WZ = 8 cm.

4. The base of a pyramid is a horizontal rectangle ABCD. The vertex E is vertically above A.
   AB = 15 m, BC = 16 m and AE = 12 m

5.

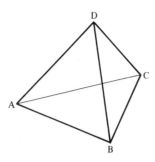

The base of the pyramid in △ABC whose area is 52 cm². The height of the pyramid is 6.8 cm.

6.

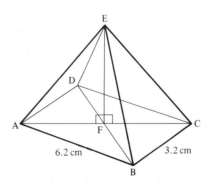

The base of the pyramid is a rectangle ABCD.
AB = 6.2 cm, BC = 3.2 cm and EF = 5.8 cm

7. The base of a pyramid is a horizontal square PQRS. The diagonals of the square meet at T. The vertex U is vertically above T. PQ = 8 cm and PU = 12 cm.

8.

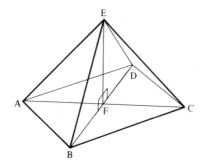

The base ABCD of the pyramid is a rectangle.
AB = 6 cm, BC = 8 cm and EC = 13 cm.
Find   a) AC and FC
   b) the height of the pyramid
   c) the volume of the pyramid.

9.

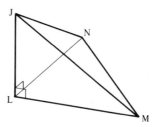

The base of the pyramid is triangle LMN.
$N\hat{L}M = 90°$, LN = 11 cm, LM = 12 cm and $J\hat{M}L = 32°$
Find   a) JL
   b) the volume of the pyramid.

MASS

When we use a weighing machine such as a spring balance we measure the *weight* of a body, that is, the pull of the earth on the body.

This weight is proportional to the amount of material in the body, i.e. to the *mass* of the body. Scientists deal more often with the mass than with the weight.

We often make use in calculations of the mass of one unit of volume of the material of which a body is made.

For instance we may be told that the mass of 1 cm^3 of silver is 10.5 g.

This is sometimes called the *density* of the material, i.e. the density of silver is 10.5 g/cm^3.

EXERCISE 3d

> The volume of a cuboid of brass is $14.3\,\text{cm}^3$. The mass of $1\,\text{cm}^3$ of brass is $8.5\,\text{g}$. Find the mass of the cuboid.
>
> $$\text{Volume} = 14.3\,\text{cm}^3$$
> $$\text{Mass} \;\;= 14.3 \times 8.5\,\text{g}$$
> $$= 121.55\,\text{g}$$
> $$= 122\,\text{g correct to 3 s.f.}$$

Find the masses of the following objects. Check first that the units are consistent.

1. A block of wood of volume $105\,\text{cm}^3$. The mass of $1\,\text{cm}^3$ of the wood is $0.68\,\text{g}$.

2. The volume of aluminium used in making a saucepan is $40\,\text{cm}^3$. The mass of $1\,\text{cm}^3$ of aluminium is $2.65\,\text{g}$.

3. A cuboid of platinum measuring $0.6\,\text{m}$ by $4.8\,\text{cm}$ by $3.2\,\text{cm}$. The mass of $1\,\text{cm}^3$ of platinum is $21.5\,\text{g}$. Give your answer correct to 3 significant figures in   a) grams   b) kilograms.

4. A litre of milk. The mass of $1\,\text{cm}^3$ of milk is $0.98\,\text{g}$.

5. An ingot of gold of volume $32\,\text{cm}^3$. The density of the gold is $19.3\,\text{g/cm}^3$.

6. A cuboid of ice measuring $4\,\text{cm}$ by $15\,\text{mm}$ by $25\,\text{mm}$. The density of ice is $0.92\,\text{g/cm}^3$.

> Find the mass of $1\,\text{cm}^3$ of oak if a block of oak measuring $3\,\text{cm}$ by $12.8\,\text{cm}$ by $5\,\text{cm}$ has a mass of $153.6\,\text{g}$
>
> $$\text{Volume} = 3 \times 12.8 \times 5\,\text{cm}^3$$
> $$= 192\,\text{cm}^3$$
> $$192\,\text{cm}^3 \text{ has mass } 153.6\,\text{g}$$
> $$1\,\text{cm}^3 \text{ has mass } \frac{153.6}{192}\,\text{g}$$
> $$= 0.8\,\text{g}$$

Find the mass of 1 cm³ of the material referred to in each of questions 7 to 10.

7. A gold cup is made from 15 cm³ of gold. The mass of the cup is 258 g.

8. The mass of two litres of petrol is 1380 g.

9. An ingot of copper in the form of a cuboid measuring 5.5 cm by 3 cm by 12 cm has a mass of 1762.2 g.

10. The volume of beech in a kitchen table is 0.24 m³ and its mass is 132 kg.

Find the density of the materials of which the bodies in questions 11 and 12 are made.

11. The volume of a china figure is 104 cm³ and its mass is 260 g.

12. The volume of a slab of granite is 4200 cm³ and its mass is 10.92 kg.

13. Find the volume of a piece of metal of mass 930 g. The mass of 1 cm³ of the metal is 15 g.

14. The mass of 1 cm³ of marble is 2.8 g. Find the volume of a marble statue whose mass is 6.86 kg.

15. A pyramid of copper has a square base of side 8 cm and a height of 10.5 cm. The mass of 1 cm³ of copper is 8.9 g. Find the mass of the pyramid.

16.

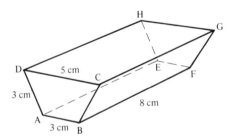

An ingot of gold is of uniform cross-section. ABCD is a trapezium with AB parallel to DC. AB = 3 cm, CD = 5 cm, BF = 8 cm and AD = BC = 3 cm.

a) Find the height of the trapezium.

b) Find the volume of the ingot.

c) The mass of 1 cm³ of the gold is 16 g. Find the mass of the ingot in
   i) grams   ii) kilograms.

d) If the gold is worth £7 per g, find the value of the ingot.

17.

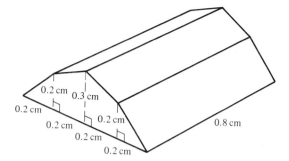

A diamond is cut in the shape of a solid of uniform cross-section as shown in the diagram.
The mass of 1 cm³ of diamond is 3.5 g.

Find   a) the volume of the diamond
       b) the mass of the diamond.

EXERCISE 3e

In this exercise several alternative answers are given. Write down the letter that corresponds to the correct answer.

1. A wooden cuboid measures 4 cm by 7 cm by 5 cm. The mass of 1 cm³ of the wood is 2.8 g.
The mass of the cuboid is

 A 50 g   **B** 392 g   **C** 44.8 g   **D** 131 g

2.

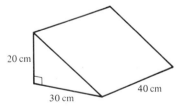

The volume of the solid of uniform cross-section is

 A 24 000 cm³   **B** 700 cm³   **C** 4800 cm³   **D** 12 000 cm³

3. The base of a pyramid is a rectangle measuring 6 cm by 10 cm. Its height is 4 cm.
Its volume is

 A 240 cm³   **B** 80 cm³   **C** 800 cm³   **D** 120 cm³

4. A cuboid of metal measuring 8 cm by 10 cm by 4 cm is melted down and made into a cuboid, two of whose measurements are 16 cm and 5 cm.
The third measurement of the new cuboid is

 A 10 cm   **B** 1 cm   **C** 4 cm   **D** 40 cm

4 FORMULAE

CONSTRUCTING FORMULAE

A formula can be thought of as a set of instructions for working out the value of a required quantity. Using this idea we can construct a formula for a given situation. For instance, to decide on the time needed to cook a turkey in foil we allow fifteen minutes per kilogram plus an extra ninety minutes.

If the turkey weighs 4 kg, then the cooking time is $(4 \times 15 + 90)$ minutes i.e. 150 minutes.

Expressing this in algebraic terms, if t minutes is the total time and the turkey weighs x kg then $t = 15x + 90$ or $t = 15(x + 6)$

If the time is T hours then we need an extra instruction to divide by 60, to change the unit from minutes to hours.

$$T = \frac{15(x + 6)}{60} \qquad \text{i.e.} \quad T = \tfrac{1}{4}(x + 6)$$

Notice that there are no units in a formula. We must make sure that the units are consistent before we start making the formula.

EXERCISE 4a

Susan is 12 years old this year. In x years time she will be y years old. Express y in terms of x.

(In, say, 5 years time, Susan will be $(12 + 5)$ years old)

$$y = 12 + x$$

1. Cakes cost 25 p each and buns cost 15 p each.

a) Find the cost in pence of 2 cakes and 3 buns.

b) Give a formula for the cost, C pence, of x cakes and y buns.

2. a) The two base angles of an isosceles triangle are 70° each. Find the third angle of the triangle.

b) The two base angles of an isosceles triangle are $x°$ each. If the third angle is $y°$ find a formula for y in terms of x.

In questions 3 to 12, consider if necessary a numerical version, then give a formula connecting the given letters.

3. A number n is equal to the mean of two numbers a and b.

4. P is the perimeter of a rectangle of length l and breadth b.

5. The cost of m metres of cloth at p pence per metre is £c (be careful about the units).

6. The cost, £C, of hiring a car for n days is a fixed charge of £A plus £D per day.

7. I can buy p tenpenny tickets for £q.

8. When making tea for n people, the number, T, of teaspoonfuls is given by the rule 'one per person and one for the pot'.

9. The nth term of the sequence $1, 3, 5, 7, \ldots$ is t

10. In a class of c pupils, each pupil is given 3 pieces of paper and 10 spare sheets are put on the teacher's desk. The total number of pieces of paper is b.

11. The sum of x metres and y centimetres is z metres.

12. The sum of three consecutive whole numbers, the first of which is n, is S.

USING FORMULAE

EXERCISE 4b

If $P = ab + c$, find P when $a = 6$, $b = -2$ and $c = 3$.

$$a = 6, b = -2, c = 3$$
$$P = ab + c$$
$$= 6 \times (-2) + 3$$
$$= -12 + 3$$
$$= -9$$

1. If $x = 3y + z$, find x when

    a) $y = 3$ and $z = 6$ b) $y = 3.2$ and $z = 4.8$

2. If $p = q - \frac{1}{2}r$, find p when

    a) $q = 14$ and $r = 30$ b) $q = 2.6$ and $r = 0.05$

3. If $M = \dfrac{6 + n}{l}$, find M if

    a) $l = 32$ and $n = 14$ b) $l = 1.2$ and $n = 8.4$

4. If $F = \dfrac{9C}{5} + 32$, find F when

    a) $C = 25$ b) $C = -6$

5. If $x = z(y - 3)$, find x if

    a) $y = 12$ and $z = 7$ b) $y = 7.2$ and $z = 1.8$

The volume $V \text{ cm}^3$ of a sphere of radius r cm is given by the formula $V = \frac{4}{3}\pi r^3$. If $\pi = 3.142$ and the radius is 2.4 cm, find the volume.

$$\pi = 3.142 \qquad r = 2.4$$

$$\text{Volume of sphere} = \tfrac{4}{3}\pi r^3$$

$$= \tfrac{4}{3} \times 3.142 \times 2.4^3 \text{ cm}^2$$

$$= 57.90 \text{ cm}^3$$

$$= 57.9 \text{ cm}^3 \quad \text{correct to 3 s.f.}$$

---

6. The surface area $A \text{ cm}^2$ of a sphere of radius r cm is given by the formula $A = 4\pi r^2$. If $\pi = 3.142$ and the radius is 6 cm, find the surface area.

7. The time, T seconds, of the swing of a pendulum of length l metres is given by $T = 2\pi\sqrt{\dfrac{l}{9.8}}$. Find the time of swing of a pendulum of length 3.2 m. (Use $\pi = 3.142$)

8. The curved surface area A of a cone of radius r and slant height l is given by $A = \pi r l$. Find the curved surface area of a cone of radius 7 cm and slant height 15 cm. $\left(\text{Use } \pi = \dfrac{22}{7} \right)$

9. From a point x metres above the sea it is possible to see a distance of y metres where $y = 100\sqrt{1274x}$. Find the distance that can be seen from a height of 10 m.

10. The kinetic energy, E joules, of an object of mass m kg and moving with velocity v m/s is given by $E = \frac{1}{2}mv^2$. Find the kinetic energy of an object of mass 2.4 kg and velocity 3.2 m/s.

11. An object starts with a speed of u m/s and during t seconds its speed increases steadily to v m/s. The distance, s metre, travelled in this time is given by $s = \left(\dfrac{u+v}{2}\right)t$. If it starts at 4.2 m/s and reaches 6.8 m/s in 4.2 s, how far does it travel ?

We may use a formula to find the value of a letter other than the subject.

EXERCISE 4c

> If $c = a(b+d)$, find b when $c = 20$, $a = 4$ and $d = 3$
>
> $$c = 20 \qquad a = 4 \qquad d = 3$$
> $$c = a(b+d)$$
> $$20 = 4(b+3)$$
> $$20 = 4b + 12$$
> $$8 = 4b$$
> $$2 = b$$
>
> i.e. $b = 2$

1. If $A = \frac{1}{2}(a+b)h$, find
   a) h when $A = 9$, $a = 2$ and $b = 1$
   b) a when $A = 12$, $b = 3$ and $h = 4$

2. If $A = \pi rl$, find l if $A = 396$, $r = 14$ and $\pi = \dfrac{22}{7}$

3. If $x = 4z + y$, find z when
   a) $x = 6$ and $y = 3$ b) $x = 9.8$ and $y = 2.6$

4. If $P = \dfrac{Q+4}{R}$ find

a) Q when $P = 7$ and $R = 3$ b) R when $P = 3$ and $Q = 3$

5. If $y = 12x^2z$ find z when

a) $x = 3$ and $y = 216$ b) $x = 0.5$ and $y = 4.8$

6. If $a = b(c+d)$ find

a) b when $a = 35.4$, $c = 4.2$ and $d = 1.7$

b) d when $a = 0.825$, $b = 1.5$ and $c = 0.3$

7. The total resistance, R, of two resistances r_1 and r_2 joined in parallel, is given by the formula $\dfrac{1}{R} = \dfrac{1}{r_1} + \dfrac{1}{r_2}$.

Find R if $r_1 = 4$ and $r_2 = 3$

8. A stone is thrown vertically upwards into the air with velocity u m/s. After t seconds it is s metres high where $s = ut - 5t^2$.

Find u if the height of the stone is 8 m after 4 seconds.

9. The area, A cm^2, of a trapezium is given by the formula $A = \frac{1}{2}(a+b)h$ where a cm and b cm are the lengths of the parallel sides and h cm is the distance between them.

a) The area is 55 cm^2 and the parallel sides are of lengths 4 cm and 7 cm. Find the distance between the parallel sides.

b) One of the parallel sides is of length 25 cm, the area is 270 cm^2 and the distance between the parallel sides is 12 cm. Find the length of the other parallel side.

10. The size, $a°$, of an interior angle of a regular n-sided polygon is given by the formula $a = \dfrac{180(n-2)}{n}$. Find the number of sides of a regular polygon if the size of each interior angle is 150°.

11. The cost, $£C$, per person for a coach trip of a distance of x miles when there are n people in the party is given by $C = \dfrac{30+x}{n}$.

a) If the cost per person is £4.50 when there are 40 people, how many miles are covered?

b) If the cost per person is £2.50 when 100 miles are travelled, how many people are there in the party?

CHANGING THE SUBJECT OF A FORMULA

Instead of using a formula in its original form to find the value of a letter other than the subject, we might prefer to rearrange the formula so that the letter we want becomes the subject. This is better if we have to use the formula several times.

Take as an example $I = \dfrac{PTR}{100}$. If we have to find P for several different sets of values of I, T and R then it is useful to have a formula for P in terms of the other letters.

To rearrange the formula we treat it as an equation and solve it for P.

$$I = \frac{PTR}{100}$$

Multiply each side by 100 $\qquad 100 \times I = \cancel{100} \times \dfrac{PTR}{\cancel{100}}$

$$100I = PTR$$

Divide each side by TR $\qquad \dfrac{100I}{TR} = P$

$\therefore \qquad\qquad\qquad\qquad\qquad P = \dfrac{100I}{TR}$

In several of the following exercises there are a few numerical equations to remind us of the processes to use.

ONE OPERATION

EXERCISE 4d

Find x in terms of the other letters if
a) $ax = b$ $\qquad$ b) $x - a = b$

$\qquad$ a) $\qquad\qquad\qquad\qquad ax = b$

Divide both sides by a $\qquad x = \dfrac{b}{a}$

$\qquad$ b) $\qquad\qquad\qquad\qquad x - a = b$

Add a to each side $\qquad\qquad x = b + a$

In questions 1 to 10, find x in terms of the other letters or numbers.

1. $x + 4 = 6$

2. $7 + x = 2$

3. $x + a = 9$

4. $6x = q$

5. $p + x = q$

6. $mx = n$

7. $x - e = f$

8. $g = hx$

9. $g = x - h$

10. $k = h + x$

In questions 11 to 20 make the letter in the bracket the subject of the formula.

11. $p = q + r$ (r)

12. $r = s - t$ (s)

13. $r = s + t$ (t)

14. $y = xz$ (z)

15. $l + m = n$ (m)

16. $P = QR$ (Q)

17. $2s = a + b + c$ (a)

18. $C = 2\pi r$ (r)

19. $A = lb$ (b)

20. $v = u + at$ (u)

TWO OR MORE OPERATIONS

EXERCISE 4e

Make x the subject of the formula $c = ax + b$

(If you cannot see what to do, put some numbers in, e.g. $8 = 2x + 4$)

$$c = ax + b$$

Take b from each side $c - b = ax$

Divide both sides by a $\dfrac{c - b}{a} = x$

i.e. $x = \dfrac{c - b}{a}$

In each question from 1 to 10, find x in terms of the other letters or numbers.

1. $4x + 3 = 11$

2. $2x - 1 = 7$

3. $6 - 2x = 2$

4. $2(x + 1) = 5$

5. $px + q = r$

6. $c = d + bx$

7. $ax - b = c$

8. $a - bx = c$

9. $a(x + b) = c$

10. $2 = p(x - q)$

In each question from 11 to 20, make the letter in the bracket the subject of the formula.

11. $ab - d = c$ (d)

12. $ab - d = c$ (a)

13. $p(q + r) = 1$ (q)

14. $3(P - Q) = 2$ (P)

15. $s = 7t + u$ (t)

16. $l - mn = 2$ (l)

17. $l - mn = 2$ (m)

18. $4T = 2P + Q$ (P)

19. $m - pr = mr$ (p)

20. $x(y - z) = 2$ (y)

21. If $P = 4(a + b)$
   a) find P when $a = 4.2$ and $b = 7.1$
   b) find a when $P = 60$ and $b = 4$
   c) make a the subject of the formula
   d) use the formula found in (c) to find a when $P = 60$ and $b = 4$
      Does your answer agree with (b)?

22. If $A = 3n(a + l)$
   a) find A when $n = 8$, $a = 3.2$ and $l = 39.2$
   b) find a when $A = 72$, $n = 6$ and $l = 6$
   c) make a the subject of the formula.
   d) Use the formula found in (c) to find a when $A = 72$, $n = 6$ and $l = 6$.
      Does your answer agree with (b)?

23. If $x = yz + z$
   a) find x when $y = 2.5$ and $z = 0.6$
   b) find y when $x = 36$ and $z = 24$
   c) make y the subject of the formula.
   d) Use the formula found in (c) to find y when $x = 36$ and $z = 24$.
      Does your answer agree with (b)?

COLLECTING LIKE TERMS

EXERCISE 4f

Make x the subject of the formula $ax + d = c + bx$

$$ax + d = c + bx$$

(Decide on which side to collect the x terms. In this case either side will do.)

Take bx from each side $ax - bx + d = c$

Take d from each side $ax - bx = c - d$

Take out the common factor $x(a - b) = c - d$

(This means that the number of xs is $(a - b)$)

Divide both sides by $(a - b)$ $x = \dfrac{c - d}{a - b}$

In questions 1 to 12, find x in terms of the other letters or numbers.

1. $3x + 6 = 5x + 1$ **5.** $ax = bx + c$ **9.** $a - bx = c + dx$

2. $7x - 4 = 2x - 1$ **6.** $ax - b = cx$ **10.** $a - bx = c - dx$

3. $6 + 2x = 8 - 3x$ **7.** $px - q = rx + q$ **11.** $p - qx = rx + sx$

4. $8 - 3x = 4 - x$ **8.** $s - tx = px$ **12.** $a + b = cx - d$

Make x the subject of the formula $x + ax = b$

$$x + ax = b$$
$$x(1 + a) = b$$
$$x = \frac{b}{1 + a}$$

In each question from 13 to 16, make x the subject of the formula.

13. $ax + x = c$ **15.** $x = cx - d$

14. $bx = x + 4$ **16.** $a(x + 1) = x(1 - a)$

In each question from 17 to 24, make the letter in the bracket the subject of the formula.

17. $pq = r - ps$ (p)

18. $ab + ac = d$ (a)

19. $ab - c = ad$ (a)

20. $ax + b = ay + c$ (a)

21. $pq + qr + rp = 0$ (p)

22. $a + b = ac$ (a)

23. $pr - p = qr + q$ (q)

24. $pr - p = qr + q$ (r)

In each question from 25 to 32, find x in terms of the other letters or numbers.

25. $2(x + 3) = 16$

26. $4(2x - 1) = 2x$

27. $3(x + 2) = 4(x - 1)$

28. $ax = b(x + 1)$

29. $a(x - b) = c$

30. $a(x + b) = bx$

31. $a(x + c) = b(x - c)$

32. $a(x + b) = c(x + d)$

In each question from 33 to 40, make the letter in the bracket the subject of the formula.

33. $p(q + r) = q$ (q)

34. $ab = c(b + a)$ (a)

35. $s(t + u) = t(u - s)$ (s)

36. $m(l - n) = n$ (n)

37. $p(q + r) = q$ (p)

38. $Q(R + P) = PR$ (R)

39. $ab = c(b + a)$ (b)

40. $s(t + u) = t(u - s)$ (u)

SQUARE ROOTS

If $x^2 = 25$ then x could be $+5$ or -5 because both $+5$ and -5 are square roots of 25.

We write $x = \pm5$

Only the positive square root is denoted by $\sqrt{25}$

so $\sqrt{25} = 5$

Hence, if $x^2 = a$, then $x = \pm\sqrt{a}$

In a problem we should always consider both square roots, as they may give two different correct answers to the question. Sometimes one of the square roots is meaningless; for instance if we are dealing with a problem about length we would probably find that only the positive root makes any sense.

Sometimes square roots can be found by inspection, even fractional square roots, e.g. $\sqrt{2\frac{1}{4}} = \sqrt{\frac{9}{4}} = \frac{3}{2}$. If the square roots are not obvious a calculator can be used.

CUBE ROOTS

Cube roots too can sometimes be found by inspection instead of by using a calculator. We can see that $\sqrt[3]{27}$ is 3 by guessing and testing. (We know $3 \times 3 \times 3 = 27$)

Even $\sqrt[3]{216}$ may be guessed: it is even and a multiple of 3, so try 6.

The cube root of a number is often much smaller than we might expect (e.g. $\sqrt[3]{512}$ is 8), so try small numbers first.

The cube root of a positive number is positive and of a negative number is negative. There is only one cube root of a number.

EXERCISE 4g

Find   a) $\sqrt{81}$ b) x if $x^2 = 81$ c) $\sqrt[3]{3\frac{3}{8}}$

a) $\sqrt{81} = 9$

b) $x^2 = 81$

   $x = \pm 9$

c) $\sqrt[3]{3\frac{3}{8}} = \sqrt[3]{\frac{27}{8}}$

   $= \frac{3}{2}$

   $= 1\frac{1}{2}$

Find, without using a calculator:

1. $\sqrt{64}$ **4.** $\sqrt{\frac{4}{9}}$ **7.** $\sqrt{\frac{1}{64}}$

2. $\sqrt[3]{125}$ **5.** $\sqrt{20\frac{1}{4}}$ **8.** $\sqrt[3]{1\,000\,000}$

3. $\sqrt{6\frac{1}{4}}$ **6.** $\sqrt[3]{15\frac{5}{8}}$ **9.** $\sqrt{1\,000\,000}$

Find x in questions 10 to 15.

10. $x^2 = 144$ **12.** $x^2 = 121$ **14.** $x^2 = 100$

11. $x^3 = 64$ **13.** $x^2 = 900$ **15.** $x^3 = -27$

FORMULAE INVOLVING SQUARES AND SQUARE ROOTS

EXERCISE 4h

a) Find x if $4x^2 = 9$

b) Make x the subject of the formula $ax^2 = b$

a) $4x^2 = 9$

$$x^2 = \frac{9}{4}$$

Take the square root of each side.

$$x = \pm\frac{3}{2}$$

b) $ax^2 = b$

$$x^2 = \frac{b}{a}$$

Take the square root of each side.

$$x = \pm\sqrt{\frac{b}{a}}$$

In questions 1 to 16 do not use a calculator.

Find x in terms of the other letters or numbers.

1. $6x^2 = 24$ **5.** $px^2 = q$

2. $9x^2 = 25$ **6.** $px^2 = q^2$

3. $3x^2 + 4 = 9$ **7.** $x^2 = p + q$

4. $x^2 = p$ **8.** $\dfrac{ax^2}{b} = c$

Find x in terms of a and b if

a) $a = b\sqrt{x}$ b) $a = \sqrt{bx}$

a) $a = b\sqrt{x}$ b) $a = \sqrt{bx}$

Square both sides   Square both sides

$a^2 = b^2 x$ $a^2 = bx$

$\dfrac{a^2}{b^2} = x$ $\dfrac{a^2}{b} = x$

i.e. $x = \dfrac{a^2}{b^2}$ i.e. $x = \dfrac{a^2}{b}$

Find x in terms of the other letters or numbers.

9. $\sqrt{x} = 4$ **13.** $p\sqrt{x} = q$

10. $3\sqrt{x} = 2$ **14.** $\sqrt{px} = r$

11. $\sqrt{3x} = 9$ **15.** $\sqrt{x} = p\sqrt{q}$

12. $\sqrt{x} = a$ **16.** $\sqrt{x+a} = 4$

In questions 17 to 29 make the letter in the bracket the subject of the formula.

17. $4p^2 = q$ (p) **21.** $\sqrt{A+B} = C$ (A)

18. $a = 2\sqrt{p}$ (p) **22.** $D = \sqrt{\dfrac{3h}{2}}$ (h)

19. $\sqrt{x+a} = b$ (a) **23.** $\sqrt{z} = \sqrt{a+b}$ (b)

20. $a^2 + b = c$ (a) **24.** $\sqrt{x^2 + a^2} = b$ (x)

25. If $z = 2(x^2 + y^2)$ find x when
a) $z = 26$ and $y = 2$ b) $z = 82$ and $y = -4$

26. A stone is dropped from a tower and after t seconds it has fallen s metres where $s = 5t^2$. Find how long it takes to fall 45 m.

27. If $P = \sqrt{Q + R}$
   a) find P when $Q = 6$ and $R = 10$
   b) find Q when $P = 5$ and $R = 5$
   c) make Q the subject of the formula.
   d) Use the formula found in (c) to find Q when $P = 5$ and $R = 5$
   Does your answer agree with (b)?

FRACTIONS

When solving a fractional equation or changing a formula involving fractions, it is most important that the fractions are removed *as soon as possible* by multiplying by the appropriate number or letter.

EXERCISE 4i

Make x the subject of the formula

a) $\dfrac{x}{a} + \dfrac{x}{b} = c$ b) $\dfrac{x}{a - b} = c$

a)

$$\frac{x}{a} + \frac{x}{b} = c$$

Multiply both sides by ab

$$ab \times \frac{x}{a} + ab \times \frac{x}{b} = ab \times c$$

$$bx + ax = abc$$

$$x(b + a) = abc$$

$$x = \frac{abc}{a + b}$$

b)

$$\frac{x}{a - b} = c$$

Multiply both sides by $(a - b)$

$$(a - b) \times \frac{x}{a - b} = (a - b) \times c$$

$$x = c(a - b)$$

In questions 1 to 12, find x in terms of the other letters or numbers. Remember to get rid of fractions first.

1. $\dfrac{x}{6} = 4$

2. $\dfrac{2x}{3} = 5$

3. $\dfrac{x}{2} + \dfrac{x}{3} = \dfrac{4}{3}$

4. $\dfrac{2x}{3} - \dfrac{x}{4} = 1$

5. $\dfrac{x}{p} = q$

6. $\dfrac{x}{p} - \dfrac{x}{q} = 1$

7. $\dfrac{ax}{p} = r$

8. $\dfrac{x}{p+q} = r$

9. $\dfrac{x}{a} + b = c$

10. $\dfrac{x+a}{b} = \dfrac{x-b}{a}$

11. $\dfrac{x}{a} = \dfrac{x+b}{a+b}$

12. $\dfrac{x}{a} + \dfrac{x}{b} = \dfrac{c}{a}$

In each question from 13 to 24, make the letter in the bracket the subject of the formula.

13. $I = \dfrac{PTR}{100}$ (R)

14. $A = \dfrac{n}{2}(a+l)$ (n)

15. $P = \dfrac{Q+R}{4}$ (Q)

16. $\dfrac{a}{3} = \dfrac{b}{2} - \dfrac{c}{4}$ (b)

17. $\dfrac{1}{x} = \dfrac{1}{a} + \dfrac{1}{b}$ (x)

18. $\dfrac{p}{x} + \dfrac{q}{x} + \dfrac{r}{x} = 1$ (x)

19. $\dfrac{s}{x} = \dfrac{r}{x} + t$ (x)

20. $a = 2\sqrt{\dfrac{p}{q}}$ (q)

21. $T = 2\pi\sqrt{\dfrac{l}{g}}$ (l)

22. $t = \dfrac{2Hh}{H+h}$ (H)

23. $a^2 + \dfrac{b}{X} + \dfrac{c}{bX} = 0$ (X)

24. $\dfrac{L}{M} = \dfrac{2a}{B-b}$ (B)

MIXED QUESTIONS

EXERCISE 4j

In each question, make the letter in the bracket the subject of the formula.

1. $v = u + at$ (t)

2. $A = \frac{1}{2}bh$ (h)

3. $a^2 = b^2 + c^2$ (c)

4. $A = \frac{1}{2}(a+b)h$ (h)

5. $\dfrac{1}{v} + \dfrac{1}{u} = \dfrac{1}{f}$ (f)

6. $A = \frac{1}{2}(a+b)h$ (a)

7. $s = \frac{t}{2}(u + v)$ (v)

8. $s = \frac{t}{2}(u + v)$ (t)

9. $\frac{1}{v} + \frac{1}{u} = \frac{1}{f}$ (u)

10. $v^2 = u^2 + 2as$ (a)

11. $A = \pi r^2 + \pi rh$ (h)

12. $v^2 = u^2 + 2as$ (u)

13. $v = \omega\sqrt{a^2 - x^2}$ (a)

14. $A = \pi r\sqrt{h^2 + r^2}$ (h)

15. $s = ut + \frac{1}{2}at^2$ (u)

16. $s = ut + \frac{1}{2}at^2$ (a)

17. $A = \frac{1}{2}pq \sin R$ (p)

18. $E = \frac{1}{2}m(v^2 - u^2)$ (u)

19. $T = 2\pi\sqrt{\dfrac{l}{g}}$ (g)

20. $A = P + \dfrac{PTR}{100}$ (R)

MIXED EXERCISES

EXERCISE 4k

1. If $Y = a + \dfrac{4}{c}$, find Y when $a = 6$ and $c = 5$.

2. If $T = \dfrac{d}{v - u}$, find v when $T = 2$, $d = 10$ and $u = 2$.

3. If $a = b\sqrt{c + d}$, find c in terms of a, b and d.

4. If $I = \dfrac{PTR}{100}$, find T in terms of I, P and R.

5. If $p = \sqrt{q^2 - r}$, find p when $q = -6$ and $r = -13$.

EXERCISE 4l

1. If $T = \dfrac{d}{v - u}$, find T when $v = 20$, $u = 6$ and $d = -7.7$.

2. If $p^2 = qr$, find p when $q = 9$ and $r = 4$.

3. If $T = \dfrac{d}{v - u}$, make d the subject of the formula.

4. If $4\sqrt{p} = q$, find p in terms of q.

5. If $a(b + c) = bc$, find b in terms of a and c.

EXERCISE 4m

In this exercise you are given several alternative answers. Write down the letter that corresponds to the correct answer.

1. If $P = Q - R^2$, $Q = 5$ and $R = 2$, then P is equal to

    A 9   **B** 49   **C** 1   **D** 9

2. If $z = \dfrac{x}{y}$, $z = 2.5$ and $y = 0.5$, then x is equal to

    A 5   **B** 0.2   **C** 1.25   **D** 3

3. If $cx + d = bx + a$, then x is equal to

    A $\dfrac{a-d}{c-b}$ **B** $\dfrac{b-c}{a+d}$ **C** $\dfrac{a+d}{b+c}$ **D** $\dfrac{c-b}{a-d}$

4. If $l = m\sqrt{n}$ then n is equal to

    A $\dfrac{l}{m}$ **B** $\dfrac{m^2}{l^2}$ **C** $\dfrac{l^2}{m^2}$ **D** $\dfrac{l^2}{m}$

5. If $\dfrac{1}{a} = \dfrac{1}{b} + \dfrac{1}{c}$ then a is equal to

    A $\dfrac{b+c}{bc}$ **B** $b+c$ **C** $\dfrac{1}{b+c}$ **D** $\dfrac{bc}{b+c}$

5 GRAPHS

THE PARABOLA

All the graphs in Book 3A, Exercise 14c came from equations of the form $y = ax^2 + bx + c$, and were curves of the same shape,

either 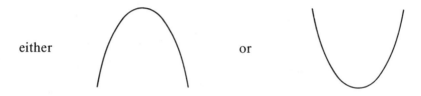 or

This shape is called a parabola. In each case there is a turning point called a vertex. When the x^2 term is *positive* the vertex is at the bottom and there is no highest value of y.

On the other hand when the x^2 term is *negative* the vertex is at the top and there is no lowest point.

The simplest equation whose graph is a parabola is $y = x^2$. Another parabola whose equation is a little more complicated than $y = x^2$ is $y = x(x - 4)$. We will plot the graph of this parabola for values of x from -1 to $+5$ at half-unit intervals.

To find the value of y for a given value of x we multiply the value of x by the value of $(x - 4)$. The table shows each value of x, the corresponding value of $x - 4$, and the resulting value of y.

x	-1	$-\frac{1}{2}$	0	$\frac{1}{2}$	1	$\frac{3}{2}$	2	$\frac{5}{2}$	3	$\frac{7}{2}$	4	$\frac{9}{2}$	5
$(x-4)$	-5	$-4\frac{1}{2}$	-4	$-3\frac{1}{2}$	-3	$-2\frac{1}{2}$	-2	$-1\frac{1}{2}$	-1	$-\frac{1}{2}$	0	$\frac{1}{2}$	1
y	5	$2\frac{1}{4}$	0	$-1\frac{3}{4}$	-3	$-3\frac{3}{4}$	-4	$-3\frac{3}{4}$	-3	$-1\frac{3}{4}$	0	$2\frac{1}{4}$	5

73

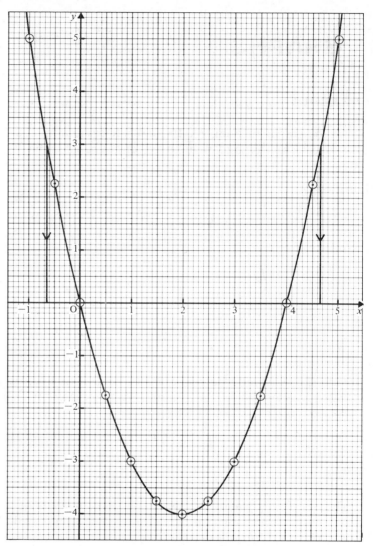

From the graph we see that the curve crosses the x-axis when $x = 0$ and when $x = 4$.

If we wish to find the value(s) of x that correspond to a given value of $x(x-4)$, i.e. a given value of y, we draw a line parallel to the x-axis for the given value of y, find where this intersects the curve, and read off the corresponding value(s) of x.

For example, if $x(x-4) = 3$, i.e. $y = 3$,
the corresponding values of x are -0.65 and 4.65.

EXERCISE 5a

1. Draw the graph of $y = 1 + 3x - x^2$ for values of x from 0 to 4 at intervals of 0.5. Use 2 cm as 1 unit on both axes.

Use your graph to find

a) the highest value of $1 + 3x - x^2$, and the corresponding value of x

b) the values of x when $1 + 3x - x^2$ has a value of   i) 0   ii) 3

2. Draw the graph of $y = x^2 - 4x + 3$ for values of x from 0 to 4 at half-unit intervals. Use 4 cm to represent 1 unit on both axes.

Use your graph to find

a) the lowest value of $x^2 - 4x + 3$, and the corresponding value of x

b) the values of x when $x^2 - 4x + 3$ has a value of   i) 2   ii) -1 iii) 4

3. Draw a graph of $y = x^2 - 4x - 9$ for values of x from -2 to $+6$ at unit intervals. Use 2 cm as 1 unit on the x-axis and 1 cm as 1 unit on the y-axis.

Use your graph to find

a) the values of x when $x^2 - 4x - 9$ has a value of -6

b) the value of $x^2 - 4x - 9$ when x is 1.4

4. Draw the graph of $y = (x - 2)(x - 3)$ for values of x from 0 to 5 at half-unit intervals. Use 2 cm as the unit on both axes.

Use your graph to find

a) the values of x when $(x - 2)(x - 3) = 0$

b) the lowest value of $(x - 2)(x - 3)$

c) the values of x for which the value of $(x - 2)(x - 3)$ is 4

5. Draw the graph of $4 + 2x - x^2$ for values of x from -2 to 4 at half-unit intervals. Use 2 cm to represent 1 unit on both axes.

Use your graph to find

a) the highest value of $4 + 2x - x^2$ and the corresponding value of x

b) the value of $4 + 2x - x^2$ when x is 2.7

c) the values of x for which the value of $4 + 2x - x^2$ is 3

6. Draw the graph of $y = (4 + x)(1 - x)$ for values of x from -5 to 2 at unit intervals. Use 2 cm to 1 unit on both axes.

Use your graph to find

a) the highest value of $(4 + x)(1 - x)$ and the corresponding value of x

b) the values of x for which $(4 + x)(1 - x) = 2$

USING A GRAPH TO SOLVE QUADRATIC EQUATIONS

From a single graph it is often possible to solve several different quadratic equations.

EXERCISE 5b

Use the graph of $y = x^2 - 2x - 4$, which is given below, to solve the equations

a) $x^2 - 2x - 4 = 0$
b) $x^2 - 2x - 4 = 6$
c) $x^2 - 2x - 2 = 0$
d) $2x^2 - 4x + 1 = 0$
e) $1 + 2x - x^2 = 0$

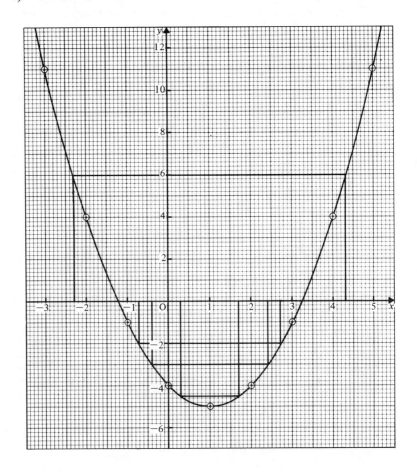

a) When this graph crosses the x-axis the value of y is 0, i.e. $x^2 - 2x - 4 = 0$. When $y = 0$ the corresponding values of x are -1.25 and 3.25. These values of x are therefore the solutions of the equation $x^2 - 2x - 4 = 0$.

b) When $x^2 - 2x - 4 = 6$, $y = 6$
The values of x that correspond to a y value of 6 are -2.3 and 4.3

Therefore the solutions of the equation $x^2 - 2x - 4 = 6$ are $x = -2.3$ and $x = 4.3$

c) To use this graph to solve the equation $x^2 - 2x - 2 = 0$ we must convert the left-hand side to $x^2 - 2x - 4$
Subtract 2 from both sides: $\quad x^2 - 2x - 4 = -2$ i.e. $y = -2$

From the graph, when $y = -2$ the values of x are -0.7 and 2.7

The solutions of the equation $x^2 - 2x - 2 = 0$ are therefore $x = -0.7$ and $x = 2.7$

(Note that the equation $x^2 - 2x - 2 = 0$ cannot be solved by factorising. The graphical solution of equations of this type is the only method available to us at the moment.)

d) To use the graph to solve $2x^2 - 4x + 1 = 0$ we must convert the LHS to $x^2 - 2x - 4$.
Divide both sides by 2 $\qquad\qquad x^2 - 2x + \tfrac{1}{2} = 0$
Subtract $4\tfrac{1}{2}$ from both sides $\qquad x^2 - 2x - 4 = -4\tfrac{1}{2}$
i.e. $\qquad\qquad\qquad\qquad\qquad y = -4\tfrac{1}{2}$
When $y = -4\tfrac{1}{2}$, $x = 0.3$ and 1.7
The solutions of the equation $2x^2 - 4x + 1 = 0$ are therefore $x = 0.4$ and $x = 1.7$

e) $1 + 2x - x^2 = 0$
(First convert the LHS to $x^2 + 2x - 4$)
Multiply both sides by -1 $\qquad -1 - 2x + x^2 = 0$
i.e. $\qquad\qquad x^2 - 2x - 1 = 0$
Subtract 3 from both sides $\qquad x^2 - 2x - 4 = -3$

From the graph, when $y = -3$, $x = 2.4$ and -0.4

The solutions of the equation $1 + 2x - x^2 = 0$ are therefore $x = 2.4$ and $x = -0.4$

1. Use the graph of $y = x^2 - 3x - 3$, which is given below, to solve the equations

a) $x^2 - 3x - 3 = 0$

b) $x^2 - 3x - 3 = 5$

c) $x^2 - 3x - 7 = 0$

d) $x^2 - 3x + 1 = 0$

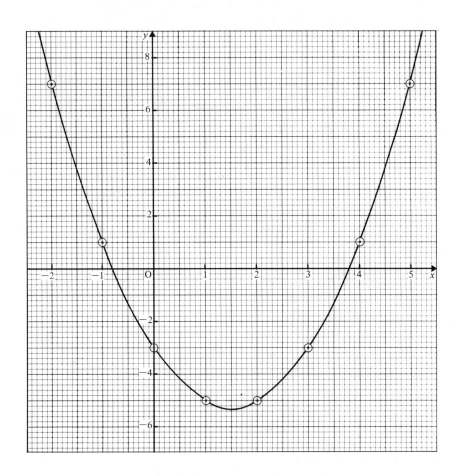

2. The graph of $y = 4 + 4x - x^2$, in the range $-1 \leqslant x \leqslant 5$, is given opposite. Use this graph to solve the equations

a) $4 + 4x - x^2 = 0$

b) $4 + 4x - x^2 = 5$

c) $4 + 4x - x^2 = 1$

d) $1 + 4x - x^2 = 0$

e) $-2 + 4x - x^2 = 0$

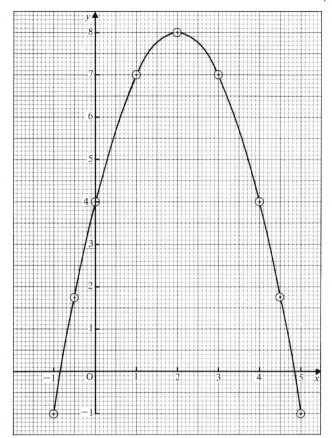

3. Use the graph of $y = x^2 - 4x + 3$, drawn for question 2 of Exercise 5a, to solve the equations

    a) $x^2 - 4x + 3 = 0$

    b) $x^2 - 4x + 1 = 0$

    c) $x^2 - 4x - 1 = 0$

    d) $x^2 - 4x - 5 = 0$

4. Use the graph of $y = 1 + 3x - x^2$, drawn for question 1 of Exercise 5a, to solve the equations

    a) $1 + 3x - x^2 = 0$

    b) $2 + 3x - x^2 = 0$

    c) $x^2 - 3x - 5 = 0$

Is it possible to use this graph to solve the equation $x^2 - 3x + 4 = 0$?
If it is possible, give the solutions. If it is not possible, explain why.

5. Draw the graph of $y = x^2 - 6x + 3$ for whole number values of x from 0 to 7 calculating the values at unit intervals. Take 2 cm as 1 unit on the x-axis and 1 cm as 1 unit on the y-axis.

Use your graph

a) to find the values of x when the graph crosses the x-axis and the equation that has these x values as solutions

b) to solve the equation $x^2 - 6x + 7 = 0$

6. Copy and complete the following table which gives values of $(2-x)(x+1)$ for values of x from -3 to 4.

x	-3	$-2\frac{1}{2}$	-2	-1	$-\frac{1}{2}$	0	$\frac{1}{2}$	1	$1\frac{1}{2}$	2	3	$3\frac{1}{2}$	4
$(2-x)$	5	$4\frac{1}{2}$		3	$2\frac{1}{2}$		$1\frac{1}{2}$	1		0	-1	$-1\frac{1}{2}$	-2
$(x+1)$	-2	$-1\frac{1}{2}$		0	$\frac{1}{2}$		$1\frac{1}{2}$	2		3	4	$4\frac{1}{2}$	5
$(2-x)(x+1)$	-10	$-6\frac{3}{4}$		0	$1\frac{1}{4}$		$2\frac{1}{4}$	2		0	-4	$-6\frac{3}{4}$	-10

Hence draw the graph of $y = (2-x)(x+1)$ for values of x from -3 to 4. Take 2 cm as 1 unit for x and 1 cm as 1 unit for y.

Use your graph to solve the equations

a) $x^2 - x - 2 = 0$

b) $x^2 - x - 5 = 0$

7. The table gives values of y for certain values of x on the curve given by the equation $y = 2x^2 - 7x + 8$.

x	0	$\frac{1}{2}$	1	$1\frac{1}{2}$	2	$2\frac{1}{2}$	3	$3\frac{1}{2}$	4
y	8	5		2	2	3		8	12

Complete this table.

[Remember that $2x^2$ means square x first and then double it.

e.g. if $x = 2\frac{1}{2}$, $y = 2 \times \frac{25}{4} - 7 \times \frac{5}{2} + 8$

$= 12\frac{1}{2} - 17\frac{1}{2} + 8$

$= 3$]

Use the table to draw the graph of $y = 2x^2 - 7x + 8$ for values of x from 0 to 4. Take 4 cm as 1 unit for x and 2 cm as 1 unit for y.

Use your graph

a) to find the lowest value of $2x^2 - 7x + 8$

b) to solve the equation $2x^2 - 7x + 4 = 0$

8. Complete the following table which shows values of $7 - 6x - 2x^2$ for values of x from -4 to 1.

x	-4	$-\frac{7}{2}$	-3	$-\frac{5}{2}$	-2	$-\frac{3}{2}$	-1	$-\frac{1}{2}$	0	$\frac{1}{2}$	1
7	7	7		7	7	7		7	7	7	
$-6x$	24	21		15	12	9		3	0	-3	
$-2x^2$	-32	$-24\frac{1}{2}$		$-12\frac{1}{2}$	-8	$-4\frac{1}{2}$		$-\frac{1}{2}$	0	$-\frac{1}{2}$	
$7 - 6x - 2x^2$	-1	$3\frac{1}{2}$		$9\frac{1}{2}$	11	$11\frac{1}{2}$		$9\frac{1}{2}$	7	$3\frac{1}{2}$	

Hence draw the graph of $y = 7 - 6x - 2x^2$ for values of x from -4 to 1. Take 4 cm as 1 unit on the x-axis and 1 cm as 1 unit on the y-axis. (If necessary turn your graph paper sideways.)

a) Use your graph to solve the equation $7 - 6x - 2x^2 = 0$

b) Draw the line $y = 5$ so that it intersects your graph. Write down the x values of the points where the line $y = 5$ meets the curve $y = 7 - 6x - 2x^2$. Find, in as simple a form as possible, the equation for which these x values are the roots.

9. Complete the following table which gives values of $3x^2 - x + 2$ for values of x from -2 to $+2$.

x	-2	-1.5	-1	-0.5	0	0.5	1	1.5	2
$3x^2$	12	6.75	3		0	0.75		6.75	
$-x$	2	1.5	1		0	-0.5		-1.5	
$+2$	2	2	2		2	2		2	
$3x^2 - x + 2$	16		6		2	2.25		7.25	

Hence draw the graph of $y = 3x^2 - x + 2$ for values of x from -2 to $+2$. Take 4 cm as 1 unit for x and 8 cm as 5 units for y.

Use your graph to solve the equations

a) $3x^2 - x - 2 = 0$

b) $3x^2 - x - 7 = 0$

c) $3x^2 - x - 1 = 0$

10. Draw a graph to solve the equation $x^2 + 2x - 4 = 0$. Take values of x from -5 to 2, and use 2 cm to represent one unit on the x-axis and 1 cm to represent one unit on the y-axis.

Hint: To solve the equation $x^2 + 2x - 4 = 0$, we need to draw the graph of $y = x^2 + 2x - 4$ and then find the values of x when $y = 0$.

11. Draw a graph to solve the equation $3 - 5x - x^2 = 0$. Take values of x from -6 to 2. Use 2 cm to represent 1 unit on the x-axis and choose your own scale for the y-axis.

12. Draw a graph to show that the equation $x^2 + 6x + 10 = 0$ cannot be solved. Take values of x from -6 to 0. Choose your own scale on each axis.

INTERSECTING GRAPHS INVOLVING QUADRATICS

In the previous exercises we have solved a quadratic equation by drawing the graph of a quadratic and finding either where the graph crossed the x-axis or where it had a particular value of y.

Sometimes it is easier to draw the simplest quadratic graph, i.e. $y = x^2$, together with a suitable straight line graph.

Suppose that we draw the graphs of $y = x^2$ and $y = 2x + 3$ on the same axes. At the points where these graphs intersect, their y values are the same. Hence the value of y on the curve (which is x^2) is equal to the value of y on the line (which is $2x + 3$) i.e., at the points of intersection of the graphs $x^2 = 2x + 3$

i.e. $x^2 - 2x - 3 = 0$

The values of x at the points of intersection must therefore be the values of x that satisfy the equation $x^2 - 2x - 3 = 0$.

This method can be used to solve a given quadratic equation. For example, to solve $x^2 - 3x - 7 = 0$ we rearrange the equation as $x^2 = 3x + 7$. Then we draw, on the same axes, the graphs of $y = x^2$ and $y = 3x + 7$.
At the points where these graphs intersect, the y values are equal,
i.e. $x^2 = 3x + 7$.

Therefore the x values at the points of intersection are the solutions of the equation $x^2 - 3x - 7 = 0$.

EXERCISE 5c

What equation can we solve by finding where the graph of $y = 5x - 2$ intersects the graph of $y = x^2$?

When the graphs intersect their y values are equal

i.e. $x^2 = 5x - 2$

or $x^2 - 5x + 2 = 0$

The values of x at the points of intersection must therefore be the solutions of the equation $x^2 - 5x + 2 = 0$.

---

1. What equations can be solved by finding where the graph of $y = x^2$ intersects the graphs of the straight lines with the following equations?

a) $y = x + 7$

b) $y = 2x + 5$

c) $y = 6x - 4$

d) $y = 5 - 3x$

2. What straight line graphs should be drawn to intersect the graph of $y = x^2$ in order to solve the following quadratic equations?

a) $x^2 - 2x - 1 = 0$

b) $x^2 - 7x + 2 = 0$

c) $x^2 + 6x + 4 = 0$

d) $2x^2 + 7x + 2 = 0$

3. *Sketch* on the same axes the graphs of $y = x^2$ and $y = x + 4$ for values of x in the range -4 to 4.
Estimate the values of x at the points of intersection of the two graphs.
What equation has these values of x as roots?

4. *Sketch* the graphs of $y = x^2$ and $y = 5x + 20$ for values of x in the range -6 to 9.
Estimate the values of x at the points where the two graphs intersect.
What equation, in its simplest form, is satisfied at these points?

See
P89

Draw the graph of $y = x^2 + 5x + 4$ for whole number values of x from -6 to 1.

Use your graph to find the lowest value of $x^2 + 5x + 4$, and the corresponding value of x.

Draw, on the same axes, the graph of $y = x + 6$. Write down the values of x at the points of intersection of the two graphs. Use your graph to find the range of values of x for which $x^2 + 5x + 4$ is less than $x + 6$. Find, in its simplest form, the equation for which the values of x at the points of intersection of the two graphs are the roots.

x	-6	-5	-4	-3	-2	-1	0	1
x^2	36	25	16	9	4	1	0	1
$5x$	-30	-25	-20	-15	-10	-5	0	5
4	4	4	4	4	4	4	4	4
$y = x^2 + 5x + 4$	10	4	0	-2	-2	0	4	10

From the graph the lowest value of $x^2 + 5x + 4$ is -2.2 which occurs when x is -2.5.

(The graph of $y = x + 6$ is a straight line so we take only three values of x and find the corresponding values of y.)

x	-6	-2	0
$y = x + 6$	0	4	6

The graphs intersect when $x = -4.45$ and 0.45.

From $x = -4.45$ to $x = 0.45$ the curve, which has equation $y = x^2 + 5x + 4$, is below the straight line, which has equation $y = x + 6$.

Therefore $x^2 + 5x + 4 < x + 6$ for all values of x greater than -4.45 but less than 0.45.

At $x = -4.45$ and $x = 0.45$ the values of y for the curve and for the straight line are equal

i.e. $x^2 + 5x + 4 = x + 6$

i.e. $x^2 + 4x - 2 = 0$

The equation $x^2 + 4x - 2 = 0$ therefore has as roots the values of x at the points of intersection of the two graphs.

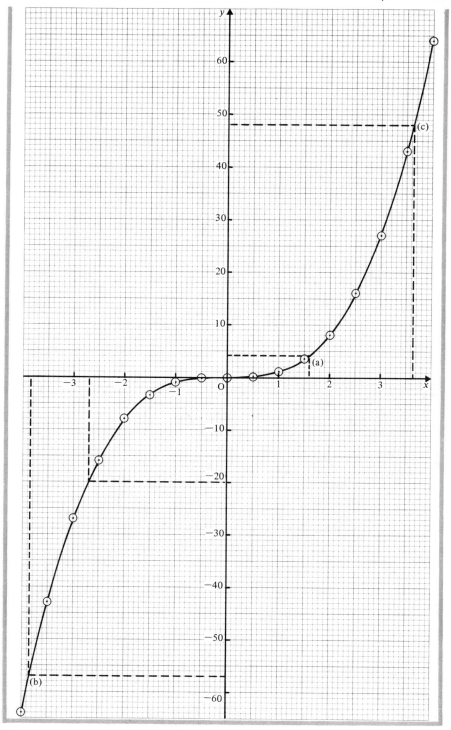

5. Complete the following table which gives values of $12 - x^2$ for values of x in the range -4 to 4.

x	-4	-3	-2	-1	0	1	2	3	4
$12 - x^2$	-4	3	8	11	12	11	8	3	-4

Using 2 cm as 1 unit for x and 1 cm as 1 unit for y, draw the graph of $y = 12 - x^2$.

Use your graph to solve the equation $12 - x^2 = 0$.

On the same axes draw the graph of $y = 2x + 7$.

What equation is satisfied by the values of x at the points of intersection of the two graphs?

Write down the values of x that satisfy this equation.

6. Copy and complete the following table which gives values of $x^2 + 2x - 2$ for values of x in the range -4 to 2.

x	-4	-3.5	-3	-2	-1	0	1	1.5	2
$x^2 + 2x - 2$	6	3.25	1	-2	-3	-2			6

Using 2 cm as unit on both axes draw the graph of $y = x^2 + 2x - 2$

On the same axes draw the graph of $y = 2 - \frac{2}{3}x$ and write down the values of x at the points where the graphs intersect.

For what range of values of x is $x^2 + 2x - 2$ less than $2 - \frac{2}{3}x$?

Find, in its simplest form, the equation for which the values of x at the intersection of the graphs, are the roots.

7. Copy and complete the following table which gives values of $9x - x^2$ for values of x in the range -2 to 10.

x	-2	-1	0	1	2	3	4	5	6	7	8	9	10
$9x - x^2$	-22		0	8	14	18	20		18	14			-10

Using 1 cm as 1 unit for x and 2 cm as 5 units for y, draw the graph of $y = 9x - x^2$.

On the same axes draw the graph of $y = 5x - 10$.

Write down the range of values of x for which $9x - x^2$ is greater than $5x - 10$.

Write down the values of x at the points of intersection of the graphs and the equation for which these values of x are the roots.

8. Write down the three values missing from the following table, which gives values of $2x^2 - 7x - 3$ for values of x in the range -1 to 5.

x	-1	-0.5	0	1	1.5	2	3	4	4.5	5
$2x^2 - 7x - 3$	6		-3	-8		-9	-6	1		12

Using 2 cm as 1 unit on the x-axis and 4 cm as 5 units on the y-axis, draw the graph of $y = 2x^2 - 7x - 3$.
Use your graph to solve the equation $2x^2 - 7x - 3 = 0$.
On the same axes draw the graph of $5x - 4y + 4 = 0$.
For what values of x is $2x^2 - 7x - 3$ less than $\frac{1}{4}(5x + 4)$?
Write down the values of x at the points of intersection of the two graphs and the equation for which these values are the roots.

9. What graph should be used, together with the graph of $y = x^2$, to solve the equation $2x^2 - x - 12 = 0$?
Sketch the two graphs for values of x in the range -4 to 4.
Use your sketch to estimate the solutions of the equation $2x^2 - x - 12 = 0$.

10. What graph should be used, together with the graph of $y = x^2$, to solve the equation $2 - 5x - x^2 = 0$?
Sketch the two graphs for values of x in the range -6 to 2.
Use your sketch to estimate the solutions of the equation $2 - 5x - x^2 = 0$.

11. Write down the three values missing from the following table, which gives values of $4x^2 - 16x + 15$ for values of x in the range 0 to 4.

x	0	0.5	1	1.5	2	2.5	3	3.5	4
$4x^2 - 16x + 15$		8	3	0	-1			8	15

Using 4 cm as 1 unit on the x-axis and 1 cm as 1 unit on the y-axis, draw the graph of $y = 4x^2 - 16x + 15$.
Use your graph to find the values of x when $y = 6$. What equation is satisfied by these values of x?
On the same axes draw the graph of $2x + y - 10 = 0$.
For what values of x is $4x^2 - 16x + 15$ less than $10 - 2x$?
Write down the values of x at the points of intersection of the two graphs and the equation for which these values are the roots.

CUBIC GRAPHS

See P85.

When an x^3 term appears in addition to the types of terms we have already considered, we obtain what is known as a cubic graph.

Examples are: $y = x^3 + x$

$$y = 2x^3 - 5$$

$$y = x^3 - 2x^2 + 3$$

and $y = 3x^3 + x^2 - 5x + 2$

The following table gives values of x^3 for values of x in the range -4 to 4.

x	-4	-3.5	-3	-2.5	-2	-1.5	-1	-0.5	0	0.5	1	1.5	2	2.5	3	3.5	4
x^3	-64	-43	-27	-16	-8	-3.4	-1	-0.1	0	0.1	1	3.4	8	16	27	43	64

These values are plotted on a graph and joined by a smooth continuous curve to give the graph shown opposite.

From the graph it is possible to find the cube of any number between -4 and $+4$ and to find the cube root of any number between -64 and $+64$.

For example, from the graph

a) $1.6^3 = 4.1$ (i.e. when $x = 1.6$, $y = 4.1$)

b) $-3.85^3 = -57$ (i.e. when $x = -3.85$, $y = -57$)

c) $\sqrt[3]{48} = 3.65$ (i.e. when $y = 48$, $x = 3.65$)

d) $\sqrt[3]{-20} = -2.7$ (i.e. when $y = -20$, $x = -2.7$)

EXERCISE 5d

1. Copy and complete the following table which gives values of $\frac{1}{5}x^3$, correct to one decimal place, for values of x from -3 to $+3$.

x	-3	-2.5	-2	-1.5	-1	-0.5	0	0.5	1	1.5	2	2.5	3
x^3	-27	-15.6	-8	-3.4		-0.13		0.13	1				27
$\frac{1}{5}x^3$	-5.4	-3.1	-1.6	-0.7		-0.03		0.03	0.2				5.4

Hence draw the graph of $y = \frac{1}{5}x^3$ for values of x from -3 to $+3$. Take 2 cm as unit on each axis.

Use your graph to solve the equations

a) $\frac{1}{5}x^3 = 4$ b) $x^3 = -15$

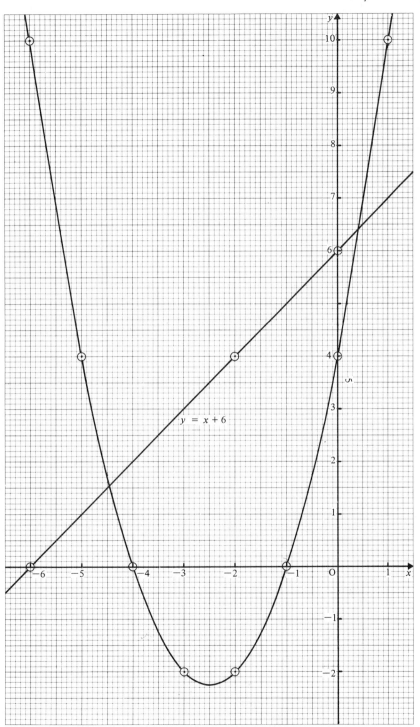

$y = x + 6$

2. Copy and complete the following table which gives values of $\frac{1}{10}x^3$ for values of x from −4 to +4.

x	−4	−3.5	−3	−2.5	−2	−1	0	1	2	2.5	3	3.5	4
x^3	−64	−42.9	−27	−15.6		−1	0	1	8			42.9	64
$\frac{1}{10}x^3$	−6.4	−4.3	−2.7	−1.6		−0.1	0	0.1	0.8			4.3	−6.4

Hence draw the graph of $y = \frac{1}{10}x^3$ for values of x from −4 to +4 taking 2 cm as unit on both axes.

Use your graph to solve the equations

a) $\frac{1}{10}x^3 = 5$ b) $x^3 = -40$

Complete the following table which gives the values of $(x+1)(x-1)(x-3)$ for values of x between −2 and +4.

x	−2	−1.5	−1	−0.5	0	0.5	1
$(x+1)(x-1)(x-3)$	−15	−5.6	0	2.6	3	1.9	

x	1.5	2	2.5	3	3.5	4
$(x+1)(x-1)(x-3)$	−1.9	−3			5.6	15

Hence draw the graph of $y = (x+1)(x-1)(x-3)$ for values of x between −2 and +4.

Use your graph to find

a) the value(s) of x when   i) $y = 1$
                              ii) $y = -8$

b) the range of values of x for which $(x+1)(x-1)(x-3)$ is positive.

(The missing values are found by substituting in the expression $(x+1)(x-1)(x-3)$.)

x	1	2.5	3
$x+1$	2	3.5	4
$x-1$	0	1.5	2
$x-3$	-2	-0.5	0
$(x+1)(x-1)(x-3)$	0	-2.625	0

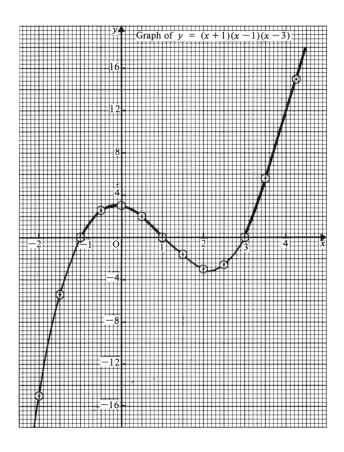

Graph of $y = (x+1)(x-1)(x-3)$

a) From the graph   i) if $y = 1$, $x = -0.9$, 0.75 3.1

                     ii) if $y = -8$, $x = -1.6$

b) If $(x+1)(x-1)(x-3)$ is positive, then y is positive. The part
of the graph for which $y > 0$ lies above the x-axis and, for
these sections, x either lies between -1 and $+1$, or is greater
than 3.

These sections are shown on the graph with a heavy line.

3. Copy and complete the table which gives the values of y when
$y = x(x-2)(x-4)$ for values of x from 0 to 4.

x	0	0.5	1	1.5	2	2.5	3	3.5	4
$x-2$		−1.5						1.5	
$x-4$		−3.5						−0.5	
y		2.265						−2.265	

Hence draw the graph of $y = x(x-2)(x-4)$, using 4 cm for 1 unit on each axis.

Use your graph to find   a) the lowest value   b) the highest value of $x(x-2)(x-4)$ within the given range of values for x.

4. Copy and complete the table which gives the value of $\frac{1}{3}x^3 - 2x + 3$ for values of x from −2 to 2.

x	−2	−1.5	−1	−0.5	0	0.5	1	1.5	2
$\frac{1}{3}x^3$	−2.67	−1.13	−0.33		0		0.33	1.13	
$-2x$	4	3	2		0		−2	−3	
$+3$	3	3	3		3		3	3	
$\frac{1}{3}x^3 - 2x + 3$	4.33	4.87	4.67		3		1.33	1.13	

Hence draw the graph of $y = \frac{1}{3}x^3 - 2x + 3$ using 4 cm for 1 unit on each axis.
Estimate the value(s) of x where the graph crosses the x-axis.

5. Copy and complete the table which gives the value of $1 - x + 2x^2 - x^3$ for values of x from −1 to 3.

x	−1	−0.5	0	0.5	1	1.5	2	3
$1-x$	2	1.5		0.5		−0.5		−2
$2x^2$	2	0.5		0.5		4.5		18
$-x^3$	1	0.125		−1.125		−3.375		−27
$1 - x + 2x^2 - x^3$	5	2.125				0.625		−11

Hence draw the graph of $y = 1 - x + 2x^2 - x^3$ using 1 cm for 1 unit on the y-axis and 2 cm for 1 unit on the x-axis.
Write down the value(s) of x where the graph crosses the x-axis.

6. *Sketch* the graph of $y = x^3$ for values of x from −2 to 2.

a) What line would you have to draw to use the graph to solve the equation $x^3 = x$?

b) Use your sketch to estimate the solution(s) of $x^3 = x$.

THE GRAPH OF $y = \dfrac{a}{x}$ WHERE a IS A CONSTANT ⎯⎯⎯⎯⎯⎯⎯

EXERCISE 5e

1. Draw a graph of $y = \dfrac{2}{x}$ for values of x from -4 to -1 and from 1 to 4. Use 2 cm as unit on both axes.

How many lines of symmetry does this curve have? Show any lines of symmetry on your diagram. Why is there no point on the graph when $x = 0$?

Use your graph to find

a) the value of y when x is 2.6

b) the value of x when y is -3.2

2. Draw the graph of $y = \dfrac{12}{x}$ for values of x from 1 to 12. Use 1 cm as unit on both axes.

Use your graph to find

a) the value of x when y is 4.6

b) the range of x values for which y is smaller than 5.6

c) the lowest value of y within the given range and the value of x for which it occurs.

3. Draw the graph of $y = \dfrac{16}{x}$, taking unit intervals for x in the range 1 to 16. Take 1 cm as unit on both axes. Draw on the same axes the graph of $x + y = 12$.

a) Write down the value(s) of x at the point(s) where the graphs intersect.

b) What equation has these x values as its roots?

c) Write down the range of values of x for which $12 - x \geqslant \dfrac{16}{x}$

4. Draw the graph of $y = \dfrac{8}{x}$ for values of x in the range -8 to -1 at unit intervals. Take 2 cm as unit on both axes. Draw on the same axes the graph of $y = x + 2$

a) Write down the value of x at the point where these graphs intersect.

b) What equation has this value as *one* of its roots? How many roots would you expect this equation to have? Explain how you could obtain the other root.

5. *Sketch* the graph of $y = \dfrac{1}{x}$ for values of x from -10 to $-\frac{1}{10}$ and $\frac{1}{10}$ to 10.

a) What happens to values of y as the value of x increases beyond $x = 10$?

b) Is there a value of y for which $x = 0$?

c) Is there a value of x for which $y = 0$?

6. Write down the three values missing from the following table, which gives values of $\dfrac{4}{x} + 1$, correct to two decimal places, for values of x in the range 0.5 to 8.

x	0.5	0.8	1	1.5	2	2.5	3	3.5	4	5	6	7	8
$\dfrac{4}{x} + 1$		6		3.67	3	2.6	2.33	2.14	2	1.8		1.57	1.5

Using the same axes, draw the graphs of $y = \dfrac{4}{x} + 1$ and $y = 4(2 - \dfrac{x}{3})$, for values of x from 0.5 to 8 taking 2 cm as 1 unit on both axes.

Use your graphs to write down the range of values of x for which $\dfrac{4}{x} + 1$ is less than $8 - \dfrac{4x}{3}$.

Write down, and simplify, the equation which is satisfied by the values of x at the points of intersection of the two graphs.

7. Sketch the graph of $y = \dfrac{10}{x}$ for values of x from 1 to 100.

a) What line has to be sketched to solve the equation $\dfrac{10}{x} = x + 1$?

b) Sketch the line on the same set of axes.

c) Does the value of x at the intersection of the graphs give *all* the solutions of $\dfrac{10}{x} = x + 1$?

EXERCISE 5f

In this exercise several possible answers are given. Write down the letter that corresponds to the correct answer.

1.

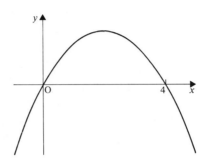

The equation of this curve could be

A $y = x^2$ **B** $y = x^3$ **C** $y = \dfrac{1}{x}$ **D** $y = 4x - x^2$

2.

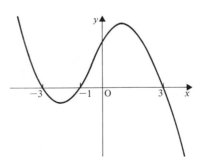

The equation of this curve could be

A $y = x^2 + x - 9$ **B** $y = (x - 3)(x + 3)(x + 1)$

C $y = \dfrac{9}{x}$ **D** $y = x^3$

3.

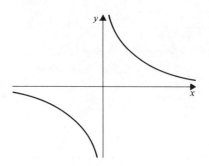

The equation of this curve could be

A $y = \dfrac{12}{x}$ **B** $y = x^2 - 9$ **C** $y = 9 - x^2$ **D** $y = x^3$

4.

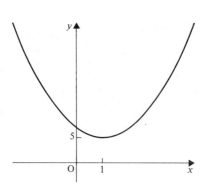

The equation of this curve could be

A $y = x^2$ **B** $y = 4 - x^2$

C $y = x^2 - 2x + 6$ **D** $y = x^3 - 4x^2 + 3$

5.

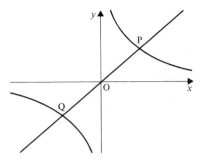

The values of x at P and Q could be the solutions of the equation.

A $x = x^2$ **B** $x = \dfrac{1}{x}$ **C** $x^2 = \dfrac{1}{x}$ **D** $x^3 = x$

6.

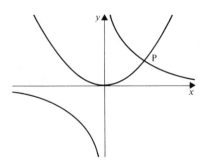

The value of x at P could be the solution of the equation

A $x^2 = \dfrac{1}{x}$ **B** $x^2 = \dfrac{1}{x^2}$ **C** $\dfrac{1}{x} = 1 + x$ **D** $x^3 = \dfrac{1}{x}$

6 INDICES

POSITIVE AND NEGATIVE INDICES

In the term a^4, 4 is called the index or power and a^4 means $a \times a \times a \times a$

We can multiply one number to a power by the *same* number to another power by adding its powers. We can divide one number to a power by the same number to another power by subtracting the powers,

e.g. $a^3 \times a^2 = a^{3+2} = a^5$ and $a^3 \div a^2 = a^{3-2} = a^1 = a$

Using the rule for division gives meaning to negative powers,

e.g. $a^3 \div a^5 = a^{-2}$ but $a^3 \div a^5 = \dfrac{a \times a \times a}{a \times a \times a \times a \times a} = \dfrac{1}{a^2}$

i.e. a^{-2} means $\dfrac{1}{a^2}$

Hence a negative sign in front of the index means 'the reciprocal of',

i.e. $$a^{-b} = \frac{1}{a^b}$$

Also, using the rule for division we have $a^2 \div a^2 = a^0$
but $$a^2 \div a^2 = 1$$
i.e. $$(\text{any number})^0 = 1$$

EXERCISE 6a

Simplify   a) $2^2 \times 2^3$ b) $3^5 \div 3^7$ c) $(0.5)^{-3}$

a) $2^2 \times 2^3 = 2^5$
$= 32$

b) $3^5 \div 3^7 = 3^{-2}$
$= \dfrac{1}{3^2}$
$= \dfrac{1}{9}$

c) $(0.5)^{-3} = \left(\frac{1}{2}\right)^{-3}$

$= \left(\frac{2}{1}\right)^{3}$

$= 8$

Simplify:

1. $3^2 \times 3^2$

2. $(2^2)^2$

3. $2^4 \times 3^2$

4. $8^5 \div 8^3$

5. $12^4 \div 12^2$

6. $7^2 \div 5^2$

7. 2^{-1}

8. $\left(\frac{1}{6}\right)^{-1}$

9. $(0.5)^{-2}$

10. $(5)^0$

11. $(1.6)^0$

12. $\left(\frac{3}{4}\right)^{-1}$

13. $2^6 \div 4^2$

14. $\left(\frac{1}{3}\right)^{-3}$

15. 2^5

16. 3^3

17. $(0.04)^{-1}$

18. $\left(\frac{3}{4}\right)^{2}$

19. $\left(\frac{2}{5}\right)^{-2}$

20. $\left(\frac{2}{3}\right)^{3}$

21. $\left(\frac{1}{6}\right)^{0}$

22. $\left(\frac{3}{5}\right)^{3}$

23. $\left(\frac{2}{5}\right)^{-1}$

24. $(0.125)^{-2}$

Simplify   a) $a^2 \div a^5$ b) $\left(\dfrac{x}{y}\right)^{-2}$

a) $a^2 \div a^5 = a^{-3}$

$= \dfrac{1}{a^3}$

b) $\left(\dfrac{x}{y}\right)^{-2} = \left(\dfrac{y}{x}\right)^{2}$

$= \dfrac{y^2}{x^2}$

Simplify:

25. $b^2 \times b^3$ **29.** $b^3 \times b^4$ **33.** $\left(\dfrac{1}{x}\right)^{-2}$

26. $c^7 \div c^5$ **30.** $\left(\dfrac{a}{b}\right)^{-1}$ **34.** $\left(\dfrac{d}{2}\right)^{-2}$

27. $\left(\dfrac{1}{c}\right)^{-2}$ **31.** $b \times b^3$ **35.** $(x^3)^{-1}$

28. $x^3 \div x^7$ **32.** $y^3 \div y$ **36.** $(p^{-1})^2$

Simplify   a) $(2^3)^2$ b) $(x^3)^2$

     a) $(2^3)^2 = 8^2$

                       $= 64$

     b) $(x^3)^2 = x^3 \times x^3$

                       $= x^6$

Simplify:

37. $(3^2)^2$ **41.** $(a^5)^2$ **45.** $(4^2)^2$

38. $(2^2)^3$ **42.** $(x^2)^3$ **46.** $(x^3)^5$

39. $(5^2)^3$ **43.** $(2^3)^3$ **47.** $(y^2)^4$

40. $(x^2)^4$ **44.** $(3^2)^3$ **48.** $(x^3)^{-2}$

Simplify $15x^5 \div 3x^3$

$$15x^5 \div 3x^3 = \frac{15x^5}{3x^3}$$

$$= 5x^2$$

Simplify:

49. $8a^3 \times 2a^2$ **53.** $6x^2 \div 3x$ **57.** $24a^2 \div 6a^4$

50. $8p^3 \div 2p^2$ **54.** $5a^2 \div 10a^3$ **58.** $3x^3 \times 4x^2$

51. $4x \div 2x^3$ **55.** $12x^6 \div 3x^2$ **59.** $8y \times 3y^3$

52. $6x^2 \times 3x$ **56.** $5y \times 3y^2$ **60.** $30y^3 \div 6y^5$

STANDARD FORM (SCIENTIFIC NOTATION)

A number between 1 and 10 multiplied by the appropriate power of 10 is said to be in standard form,

e.g.   when 0.0043 is written in standard form
it becomes 4.3×10^{-3}

EXERCISE 6b

Write in standard form   a) 3700   b) 0.052

a) $3700 = 3.7 \times 10^3$

b) $0.052 = 5.2 \times 10^{-2}$

Write the following numbers in standard form.

1. 280   **4.** 0.097   **7.** 0.8

2. 0.39   **5.** 2770   **8.** 8000

3. 707   **6.** 0.00008   **9.** 0.025

If $a = 1.2 \times 10^{-2}$ and $b = 6 \times 10^{-4}$ express in standard form

a) ab b) $\dfrac{a}{b}$ c) $a + b$

If $a = 1.2 \times 10^{-2}$ and $b = 6 \times 10^{-4}$ then

a) $ab = 1.2 \times 10^{-2} \times 6 \times 10^{-4}$

$\qquad = 7.2 \times 10^{-6}$

b) $a = \dfrac{1.2 \times 10^{-2}}{6 \times 10^{-4}}$

$\qquad b = \dfrac{1.2}{6} \times 10^{-2-(-4)}$

$\qquad = 0.2 \times 10^{2}$

(This number must now be converted to standard form)

$\qquad = 2.0 \times 10^{-1} \times 10^{2}$

$\qquad = 2 \times 10^{1}$

c) $a + b = 1.2 \times 10^{-2} + 6 \times 10^{-4}$

(Multiplication must be done before addition, so each number must be written in full.)

$a + b = 0.012 + 0.0006$

$\qquad = 0.0126$

$\qquad = 1.26 \times 10^{-2}$

Write down the value of ab in standard form if

10. $a = 2.1 \times 10^{2}$, $b = 4 \times 10^{3}$

11. $a = 5.4. \times 10^{4}$, $b = 2 \times 10^{5}$

12. $a = 7 \times 10^{-2}$, $b = 2.2 \times 10^{-3}$

13. $a = 5 \times 10^{-4}$, $b = 2.3 \times 10^{-2}$

14. $a = 1.6 \times 10^{-2}$, $b = 2 \times 10^{4}$

15. $a = 6 \times 10^{5}$, $b = 1.3 \times 10^{-7}$

Write down the value of $\frac{p}{q}$ in standard form if

16. $p = 6 \times 10^{-5}$, $q = 3 \times 10^2$

17. $p = 1.4 \times 10^8$, $q = 2 \times 10^3$

18. $p = 9 \times 10^3$, $q = 3 \times 10^5$

19. $p = 7 \times 10^{-3}$, $q = 5 \times 10^2$

20. $p = 1.8 \times 10^{-3}$, $q = 6 \times 10^{-4}$

21. $p = 2.5 \times 10^4$, $q = 2 \times 10^{-4}$

Write down the value of $x + y$ in standard form if

22. $x = 2 \times 10^2$, $y = 3 \times 10^3$

23. $x = 3 \times 10^{-2}$, $y = 2 \times 10^{-3}$

24. $x = 2.1 \times 10^4$, $y = 3.1 \times 10^5$

25. $x = 1.3 \times 10^{-4}$, $y = 4 \times 10^{-3}$

26. $x = 1.9 \times 10^{-3}$, $y = 2.4 \times 10^{-2}$

27. $x = 3 \times 10^5$, $y = 2.5 \times 10^6$

28. If $x = 1.2 \times 10^5$ and $y = 5 \times 10^{-2}$ find, in standard form, the value of

a) xy b) $x \div y$ c) $x + 1000y$

29. If $m = 7.2 \times 10^{-7}$ and $n = 1.2 \times 10^{-5}$ find, in standard form, the value of

a) mn b) $m \div n$ c) $n - m$

30. If $u = 2.6 \times 10^5$ and $v = 5 \times 10^{-3}$ find, in standard form, the value of

a) uv b) $u \div v$ c) $\dfrac{u}{100} + 100v$ d) $\dfrac{u}{10} - 1000v$

FRACTIONAL INDICES

Consider $a^{1/2}$.

To give a meaning to $a^{1/2}$ we use the fact that

$$a^{1/2} \times a^{1/2} = a^{1/2 + 1/2} = a^1$$

Now when one number is multiplied by itself, the result is called the square of this number,

i.e. a is the *square* of $a^{1/2}$.

Taking $a^{1/2}$ as a positive number, this means that

$a^{1/2}$ is the *positive square root* of a

Hence $4^{1/2}$ means 'the positive square root of 4'

i.e.
$$4^{1/2} = \sqrt{4} = 2$$

Similarly, $a^{1/3} \times a^{1/3} \times a^{1/3} = a^{1/3+1/3+1/3} = a^1$ so that $a^{1/3}$ is the *cube root* of *a*

Hence $27^{1/3}$ means 'the cube root of 27',

i.e. $$27^{1/3} = \sqrt[3]{27} = 3$$

In the same way $a^{1/4}$ means 'the positive fourth root of *a*' so that

$$16^{1/4} = \sqrt[4]{16} = 2$$

(since $2 \times 2 \times 2 \times 2 = 16$)

In general $a^{1/n}$ means 'the *n* th root of *a*',
i.e. $a^{1/n}$ is the number which, when *n* of them are multiplied together, gives *a*.

When a number can have both a positive and a negative root, $a^{1/n}$ means the positive root only.
e.g. both 2 and -2 are square roots of 4
    but $4^{1/2} = 2$

EXERCISE 6c

Simplify:

1. $9^{1/2}$

2. $16^{1/2}$

3. $36^{1/2}$

4. $8^{1/3}$

5. $125^{1/3}$

6. $64^{1/3}$

7. $\left(\frac{1}{4}\right)^{1/2}$

8. $(0.04)^{1/2}$

9. $\left(\frac{1}{8}\right)^{1/3}$

10. $\left(\frac{4}{9}\right)^{1/2}$

11. $(0.25)^{1/2}$

12. $\left(\frac{8}{27}\right)^{1/3}$

Consider $8^{2/3}$.

$$8^{2/3} = 8^{1/3} \times 8^{1/3} = (8^{1/3})^2$$

Therefore $8^{2/3}$ can be read as 'the square of the cube root of 8', i.e. $(\sqrt[3]{8})^2$

Alternatively, since $8^{2/3} \times 8^{2/3} \times 8^{2/3} = 8^2$, $8^{2/3}$ can be read as 'the cube root of 8 squared', i.e. $\sqrt[3]{(8^2)}$

Therefore $8^{2/3}$ can be evaluated by

either   1) finding the cube root of 8 and squaring the result
        i.e. $8^{2/3} = (8^{1/3})^2 = 2^2 = 4$
or   2) squaring 8 and finding the cube root of the result
        i.e. $8^{2/3} = (8^2)^{1/3} = 64^{1/3} = 4$

Finding the required root first (method 1) keeps the size of the numbers down, so this should be used whenever possible.

EXERCISE 6d

Simplify   a) $4^{3/2}$ b) $\left(\frac{1}{4}\right)^{-1/2}$

a) $4^{3/2} = (\sqrt{4})^3 = 2^3 = 8$

b) $\left(\frac{1}{4}\right)^{-1/2} = (4)^{1/2} = 2$

Simplify:

1. $(27)^{2/3}$

2. $\left(\frac{1}{8}\right)^{2/3}$

3. $(16)^{3/4}$

4. $(125)^{2/3}$

5. $(0.008)^{2/3}$

6. $(144)^{3/2}$

7. $(0.36)^{3/2}$

8. $81^{3/4}$

9. $32^{2/5}$

10. $(1000)^{2/3}$

11. $(0.0001)^{3/4}$

12. $(100\,000)^{2/5}$

13. $\left(\frac{1}{9}\right)^{-1/2}$

14. $\left(\frac{4}{49}\right)^{-1/2}$

15. $(0.04)^{-1/2}$

16. $\left(\frac{8}{27}\right)^{-1/3}$

17. $8^{-2/3}$

18. $(32)^{-3/5}$

19. $(16)^{-1/4}$

20. $(0.01)^{-3/2}$

21. $(1000)^{-2/3}$

22. $\left(\frac{1}{9}\right)^{-3/2}$

23. $(0.027)^{-2/3}$

24. $(6.25)^{-1/2}$

25. $(x^2)^{1/4}$

26. $(x^{1/3})^6$

27. $(y^4)^{1/2}$

28. $(a^6)^{1/3}$

29. $(x^8)^{3/4}$

30. $(x^{2/3})^6$

USING A CALCULATOR

To find the cube root of 12 using a calculator, first write $\sqrt[3]{12}$ as $12^{1/3}$ then

a) if your calculator has a key $\boxed{y^{1/x}}$, use the following sequence:

$\boxed{1}$ $\boxed{2}$ $\boxed{y^{1/x}}$ $\boxed{3}$ $\boxed{=}$

b) if your calculator has a key $\boxed{y^x}$ but not $\boxed{y^{1/x}}$, change the fractional index to a decimal, i.e.

$\boxed{1}$ $\boxed{2}$ $\boxed{y^x}$ $\boxed{.}$ $\boxed{3}$ $\boxed{3}$ $\boxed{3}$ $\boxed{3}$ $\boxed{=}$

Any root can be found in a similar manner,

e.g. $\sqrt[4]{20} = 20^{1/4} = 20^{0.25}$ and $(0.27)^{2/3} = (0.27)^{0.667}$

EXERCISE 6e

Use your calculator to evaluate the following roots, giving your answers correct to 3 s.f.

1. $\sqrt[3]{24}$

2. $(24)^{1/2}$

3. $\sqrt[4]{100}$

4. $\sqrt[5]{216}$

5. $\sqrt[3]{0.01}$

6. $(1.8)^{2/3}$

7. $\sqrt[3]{502}$

8. $\sqrt[4]{36}$

9. $(0.2)^{3/5}$

10. $(1.5)^{1/5}$

11. $\sqrt[6]{0.1}$

12. $\sqrt[4]{24.2}$

MIXED EXERCISES

EXERCISE 6f

In this exercise each question is followed by several alternative answers. Write down the letter that corresponds to the correct answer.

1. The value of $\left(\frac{1}{10}\right)^{-3}$ is

    A $\frac{1}{30}$ **B** $\frac{1}{1000}$ **C** 1000 **D** 30

2. The value of $(1.6)^0$ is

    A 0 **B** 1 **C** 1.6 **D** $\frac{1}{16}$

3. When $a = 2 \times 10^2$ and $b = 3 \times 10^4$, the value of $a \times b$ is

    A 5×10^6 **B** 6×10^8 **C** -1×10^{-2} **D** 6×10^6

4. The value of $\left(\frac{27}{8}\right)^{-1/3}$ is

    A $1\frac{1}{2}$ **B** $\frac{2}{3}$ **C** $-\frac{8}{81}$ **D** $-\frac{27}{24}$

5. If $x = 4 \times 10^{-3}$ and $y = 2 \times 10^{-2}$ then the value of $x + y$ is

    A 6×10^{-5} **B** 4.2×10^{-3} **C** 8×10^{-5} **D** 2.4×10^{-2}

6. $2p^2 \times 3p^3$ can be written as

    A $5p^5$ **B** $6p^6$ **C** $6p^5$ **D** $5p^6$

7. $\left(\frac{x}{y}\right)^{-2}$ can be written as

    A $\frac{y^2}{x^2}$ **B** xy^2 **C** $\sqrt{\left(\frac{x}{y}\right)}$ **D** x^2y^2

EXERCISE 6g

1. Simplify:

   a) $\left(\frac{1}{2}\right)^{-2}$ b) $49^{-1/2}$ c) $8^{4/3}$ d) $25^{3/2}$

2. If $x = 16$ and $y = 27$ find

   a) $x^{-1/2}$ b) $y^{1/3}$ c) $(xy)^0$ d) $\left(\frac{x}{y}\right)^{-1}$

3. If $a = \frac{1}{4}$ and $b = \frac{1}{25}$ find

   a) a^2 b) b^0 c) $\left(\frac{b}{a}\right)^{1/2}$ d) $a^{1/2}b^{-1}$

4. If $u = 2.7 \times 10^4$ and $v = 3 \times 10^2$ find, in standard form, the value of
   a) uv b) $u + v$ c) $u \div v$

5. Simplify:

   a) $x^2 \times x^{-3}$ b) $4a^5 \div 2a^7$ c) $\left(\frac{x^2}{y}\right)^{-1}$

6. Find the value of x when   a) $x^3 = 8$ b) $2^x = 16$

EXERCISE 6h

1. Simplify:

   a) $\left(\frac{3}{4}\right)^2$ b) $\left(\frac{3}{5}\right)^{-1}$ c) $8^{1/3}$ d) $\left(\frac{1}{3}\right)^0$

2. If $x = 1.8 \times 10^{-3}$ and $y = 2.4 \times 10^{-2}$ find in standard form the value of

   a) xy b) $\dfrac{x}{y}$ c) $x + y$ d) $y - x$

3. If $p = 25$ and $q = 6$ find

   a) $p^{1/2}$ b) q^{-2} c) $p^{-3/2}$ d) $\left(\frac{p}{q}\right)^{-1}$

4. If $a = \frac{2}{3}$ and $b = \frac{3}{5}$ find

   a) a^{-2} b) b^2 c) $(ab)^{-1}$ d) $\left(\frac{a}{b}\right)^2$

5. Simplify:
   a) $p^{-2} \times p^4$ b) $4x^2 \div 8x^3$ c) $(x^3)^4$

6. Find the value of x when
   a) $x^4 = 81$ b) $4^x = 64$

EXERCISE 6i

1. Find:
   a) $\left(\frac{2}{3}\right)^4$ b) $\left(\frac{4}{9}\right)^{1/2}$ c) $\left(\frac{3}{7}\right)^{-2}$ d) $4^{3/2}$

2. Find:
   a) $(2^2)^3$ b) $(125)^{2/3}$ c) $32^{2/5}$ d) $\left(\frac{1}{2}\right)^{-3}$

3. Simplify:
   a) $x^5 \div x^3$ b) $12y^2 \div 8y^5$ c) $(y^2)^4$

4. If $a = 3.2 \times 10^5$ and $b = 2 \times 10^4$ find in standard form the value of
   a) ab b) $a \div b$ c) $b \div a$ d) $a - b$

5. Simplify:
   a) $(p^2)^5$ b) $(x^{1/2})^3$ c) $8y^3 \div 20y^4$

6. Find the value of x when
   a) $x^2 = \frac{4}{9}$ b) $8^x = 2$

7 CYLINDERS, CONES AND SPHERES

AREA AND CIRCUMFERENCE

In Book 2 we found that the circumference, C, of a circle of radius r is given by the formula

$$C = 2\pi r$$

where π is the symbol for a number which has no exact numerical value and for which we use the approximations 3.14 or 3.142 or the value given in your calculator or occasionally $\frac{22}{7}$.

The area, A, of a circle is given by the formula

$$A = \pi r^2$$

For all calculations in this chapter use the value of π given by your calculator unless otherwise instructed. For estimations (i.e. checks on answers) take $\pi \approx 3$.

EXERCISE 7a

Find the circumference and area of a circle of radius 6.2 m.

$$r = 6.2$$

Using $C = 2\pi r$

the circumference $= 2 \times \pi \times 6.2 \, \text{cm}$

$= 38.95 \, \text{cm}$

(Check: circumference $\approx 2 \times 3 \times 6 \, \text{cm} = 36 \, \text{cm}$)

Using $A = \pi r^2$

the area $= \pi \times (6.2)^2 \, \text{cm}^2$

$= 120.7 \, \text{cm}^2$

(Check: area $\approx 3 \times 6^2 \approx 3 \times 40 = 120 \, \text{cm}^2$)

The circumference is 39.0 cm and the area is 121 cm^2 correct to 3 s.f.

Find the circumference and area of the circle whose radius is given.

1.	9 cm	**4.**	23 mm	**7.**	1.06 cm
2.	3.2 cm	**5.**	13 cm	**8.**	14.2 mm
3.	24 m	**6.**	2 m	**9.**	2.9 cm

10.	7.3 cm	**13.**	8.8 cm	**16.**	19 cm
11.	0.9 m	**14.**	103 mm	**17.**	40.7 cm
12.	19.1 mm	**15.**	1.2 m	**18.**	93 mm

Find the area of a circle of radius 4 cm, giving your answer as a multiple of π.

$$r = 4$$

Using $A = \pi r^2$

$$\text{the area} = \pi \times (4)^2 \, \text{cm}^2$$

$$= 16\pi \, \text{cm}^2$$

In each question from 19 to 24, find the circumference and area of the circle whose radius is given. Give your answer as a multiple of π.

19.	3 cm	**21.**	80 cm	**23.**	4.5 m
20.	12 cm	**22.**	1 m	**24.**	11 mm

25. Find the circumference of a bicycle wheel of diameter 66 cm.

26. Find the area of a dartboard of diameter 44 cm.

27. A circle of radius 4 cm is cut from a square piece of paper of side 10 cm. Find the area of the remaining paper.

28. Find the perimeter and area of a semicircle of radius 18 cm.

29.

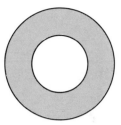

The shaded ring (sometimes called an *annulus*) is bounded by two concentric circles (i.e. circles with the same centre) of radii 10 cm and 6 cm. Find the shaded area.

30.

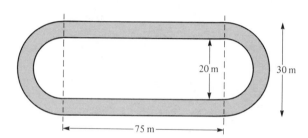

The diagram shows a running track in which the curved boundaries are semicircles.

Find   a) the outer perimeter of the track

b) the inner perimeter of the track

c) the shaded area.

31. Find the area of the annulus which is bounded by circles of radii 9 cm and 5 cm. Give your answer as a multiple of π.

In questions 32 and 33 several alternative answers are given. Write down the letter that corresponds to the correct answer.

32. The area of a circle of radius 6 cm is

A $6\pi^2 \text{cm}^2$ **B** $72\pi \text{cm}^2$ **C** $12\pi \text{cm}^2$ **D** $36\pi \text{cm}^2$

33. The circumference of a circle of radius 21 cm is approximately

A 66 cm   **B** 1400 cm   **C** 130 cm   **D** 1300 cm

INVERSE PROBLEMS

EXERCISE 7b

The circumference of a circle is 40 cm. Find its radius.

$$C = 40$$

$$C = 2\pi r$$

$$\therefore \quad 40 = 2\pi r$$

$$\frac{40}{2\pi} = r$$

$$r = 6.366$$

$$\left(\text{Check: } r \approx \frac{40}{2 \times 3} \approx 6\right)$$

The radius is 6.37 cm correct to 3 s.f.

In each question from 1 to 18 find the radius of the circle whose circumference is given.

1.	18.2 cm	**4.**	2.6 cm	**7.**	4.2 m
2.	6.8 cm	**5.**	30 m	**8.**	108 cm
3.	14 cm	**6.**	192 m	**9.**	13 mm

10.	148 mm	**13.**	3.4 m	**16.**	236 mm
11.	99 cm	**14.**	63 mm	**17.**	6.9 m
12.	14π cm	**15.**	76π mm	**18.**	90π cm

The area of a circle is $28\,\text{cm}^2$. Find its radius.

$$A = 28$$

$$A = \pi r^2$$

$$\therefore \qquad 28 = \pi r^2$$

$$\therefore \qquad \frac{28}{\pi} = r^2$$

$$\therefore \qquad r = \sqrt{\frac{28}{\pi}}$$

$$= 2.985$$

(Check: $r^2 \approx \dfrac{28}{3} \approx 9,\ r \approx 3$)

The radius is $2.99\,\text{cm}$ correct to 3 s.f.

In each question from 19 to 27 find the radius of the circle whose area is given.

19. $42\,\text{cm}^2$	**22.** $4\,\text{m}^2$	**25.** $9\,\text{cm}^2$
20. $119\,\text{m}^2$	**23.** $124\,\text{cm}^2$	**26.** $1200\,\text{mm}^2$
21. $26\,\text{mm}^2$	**24.** $56\,\text{cm}^2$	**27.** $13\,\text{m}^2$

28. Find the radius of a circle whose circumference is $220\,\text{mm}$.
What is the area of the circle?

29. Find the radius and area of a circle whose circumference is $100\,\text{cm}$.

30. Liquid spilt on a floor forms a circular patch of area $30.6\,\text{cm}^2$. What is the radius of the patch?

31. Find the radius and circumference of a circle whose area is $62\,\text{cm}^2$.

32. Find the radius of a semicircle whose area is $15\,\text{cm}^2$.

ARC AND SECTOR

Part of the circumference of a circle is called an *arc* of a circle.

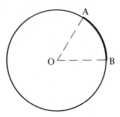

Its shape is defined by the radius of the circle and the angle it subtends at the centre, i.e. $A\hat{O}B$.

A *sector* is a slice of a circle, defined by the radius and the angle at the centre.

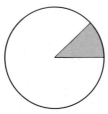

Do not confuse a sector with a *segment,* which is cut from a circle by a chord.

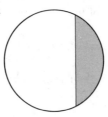

A chord divides a circle into two segments, a minor segment (shaded) and a major segment (unshaded).

EXERCISE 7c

a) Find the length of an arc that subtends an angle of 30° at the centre of a circle of radius 8 cm.

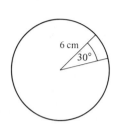

Arc length is $\dfrac{30}{360}$ of circumference

$$= \dfrac{\cancel{30}^{1}}{\cancel{360}_{12}} \times 2\pi r$$

$$= \dfrac{1}{12} \times 2 \times \pi \times 8 \,\text{cm}$$

$$= 4.189 \,\text{cm}$$

∴ the length of the arc is 4.19 cm correct to 3 s.f.

b) Find the area of the sector described in (a).

Area of sector is $\dfrac{30}{360}$ of area of circle

$$= \dfrac{\cancel{30}^{1}}{\cancel{360}_{12}} \times \pi r^2$$

$$= \dfrac{1}{12} \times \pi \times (8)^2 \,\text{cm}^2$$

$$= 16.75 \,\text{cm}^2$$

∴ the area of the sector is 16.8 cm² correct to 3 s.f.

In each question from 1 to 6 find

a) the length of the arc   b) the area of the sector.

1.

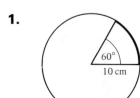

2.

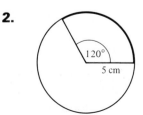

3.

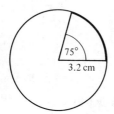

5.

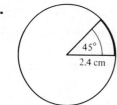

4.

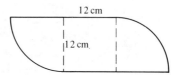

6.

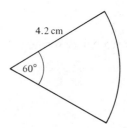

7. Find the perimeter and area of a quadrant of a circle of radius 10 cm.

8.

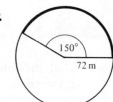

A figure is formed from a square of side 12 cm and two quadrants. Find its perimeter and area.

9.

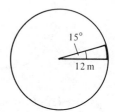

A sector of a circle of radius 4.2 cm contains an angle of 60°.

Find   a) the perimeter of the sector

        b) the area of the sector.

Find the angle at the centre of a circle of radius 10 cm subtended by an arc of length 5 cm.

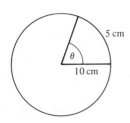

$$\frac{\text{Arc length}}{\text{Circumference}} = \frac{\theta}{360°}$$

$$\frac{5}{2\pi r} = \frac{\theta}{360°}$$

$$\frac{\overset{18}{\cancel{360}} \times 5}{\cancel{2} \times \pi \times \cancel{10}} = \theta$$

$$\theta = 28.64°$$

The angle at the centre is 28.6°

In each question from 10 to 12 find the angle subtended at the centre of the circle.

10. Arc length 7 cm, radius of circle 6 cm.

11. Arc length 60 cm, radius of circle 80 cm.

12. Arc length 3.2 cm, radius of circle 12 cm.

In each question from 13 to 15 find the radius of the circle.

13. Arc length 3 cm, angle at the centre 20°.

14. Arc length 7 cm, angle at the centre 45°.

15. Arc length 9 m, angle at the centre 150°.

16. Use $\pi = 3.142$ for this question, *not* the value of π given by a calculator.

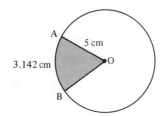

The length of the arc AB is 3.142 cm and the radius of the circle is 5 cm.

a) Find $A\widehat{O}B$.

b) Find the area of the shaded sector.

17. A cone made of cardboard has a base of radius 2 cm and slant height 6 cm.

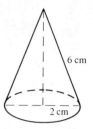

a) Find the circumference of the base circle.

b) The curved surface of the cone is flattened out into a sector of a circle. Draw a diagram and mark in its measurements.

c) Find the angle of the sector.

d) Find the area of the sector.

e) Find the total surface area of the cone.

18.

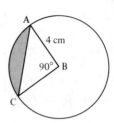

Using the data given in the diagram, find

a) the area of △ABC

b) the area of the shaded segment.

19.

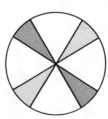

A pattern is formed by painting a white circle of radius 15 cm with four sectors of angle 27° each. Two are painted red and two blue. Find the total area of

a) the white parts

b) the red parts.

ANOTHER APPROXIMATE VALUE FOR π

If the radius of the circle is a multiple of 7, then $\frac{22}{7}$ may be used as an approximate value of π. The result is not as accurate as the one given by using the value of π from a calculator but the arithmetic may be so easy that a calculator is not needed.

EXERCISE 7d

In this exercise use $\frac{22}{7}$ for π. Do not use a calculator.

Find the circumference of a circle of radius $3\frac{1}{2}$ cm.

$$r = 3\frac{1}{2}, \quad \pi = \frac{22}{7}$$

Using $C = 2\pi r$

the circumference $= 2 \times \frac{22}{7} \times \frac{7}{2}$ cm

$= 22$ cm

In each question from 1 to 6 find the circumference and area of the circle whose radius is given.

1. 7 cm

2. $1\frac{3}{4}$ cm

3. $17\frac{1}{2}$ cm

4. 70 m

5. $\frac{7}{10}$ cm

6. $2\frac{1}{3}$ m

7. Find the perimeter of a semicircle of radius $3\frac{1}{2}$ cm.

8. Find the area of a quadrant of a circle of radius 7 cm.

9. Find the length of an arc subtending an angle of 45° at the centre of a circle of radius 28 m.

10. Find the area of a sector of a circle of radius 21 cm if the angle at the centre is 40°.

Find the radius of a circle of area $9\frac{5}{8}$ cm².

$$A = 9\frac{5}{8}, \ \pi = \frac{22}{7}$$

Using $A = \pi r^2$

$$\frac{77}{8} = \frac{22}{7} \times r^2$$

Multiply both sides by $\dfrac{7}{22}$

$$\frac{7}{{}_{2}\cancel{22}} \times \frac{\cancel{77}^{7}}{8} = \frac{\cancel{7}^{1}}{{}_{1}\cancel{22}} \times \frac{\cancel{22}^{1}}{\cancel{7}_{1}} \times r^2$$

$$\therefore \qquad r^2 = \frac{49}{16}$$

$$r = \frac{7}{4}$$

$$= 1\frac{3}{4}$$

The radius is $1\frac{3}{4}$ cm.

In each question from 11 to 16, find the radius of the circle whose area or circumference is given.

11. Circumference 44 cm

12. Circumference 176 cm

13. Area $9\frac{5}{8}$ cm²

14. Circumference 33 cm

15. Area $6\frac{4}{25}$ m²

16. Area 15 400 cm²

CYLINDERS

THE CURVED SURFACE AREA OF A CYLINDER

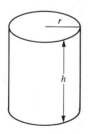

If we have a cylindrical tin with a paper label covering its curved surface, we can take off the label and flatten it out to give a rectangle whose length is equal to the circumference of the tin.

Therefore the area A of the curved surface is given by $2\pi r \times h$

i.e. $$A = 2\pi rh$$

EXERCISE 7e

In each question from 1 to 8, find the curved surface area of the cylinder whose measurements are given.

1. Radius 4 cm, height 6 cm

2. Radius 30 cm, height 2 cm

3. Radius 6.2 cm, height 5.8 cm

4. Radius 2 m, height 82 cm

5. Radius 0.06 m, height 32 cm

6. Radius 5.2 cm, height 7.8 cm

7. Radius 72.6 cm, height 30 cm

8. Radius 4.2 m, height 9.8 m.

9. A closed cylinder has radius 6 cm and height 10 cm.
   Find   a) the area of its curved surface
          b) the area of its base
          c) the total surface area.

10. A closed cylinder has radius 3.2 cm and height 4.8 cm.
   Find   a) the area of its curved surface
          b) the total surface area.

11. Find the area of the paper label covering the side of a cylindrical soup tin of height 9.6 cm and radius 3.3 cm. The label has an overlap of 1 cm.

12. What area of card is needed to make a cylindrical tube of length 42 cm and radius 3.2 cm? The card overlaps by 2 cm.

13. A garden roller is in the form of a cylinder of radius 0.25 m and width 0.7 m. In four revolutions of the roller what area of lawn does it roll?

VOLUME OF A CYLINDER

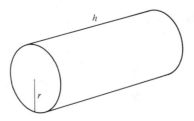

We can think of a cylinder as a solid of uniform circular cross-section. Its volume, V, is therefore found by multiplying the area of the cross-section by the length, so $V = \pi r^2 \times h$

i.e. $$V = \pi r^2 h$$

EXERCISE 7f

a) Find the volume inside a hollow cylinder of radius 9.8 cm and height 6.7 cm.
b) What is its capacity in litres?

a) $r = 9.8 \qquad h = 6.7$

$V = \pi r^2 h$

$= \pi \times (9.8)^2 \times 6.7$

$= 2021$

(Check: $V \approx 3 \times 100 \times 7 = 2100$)

The volume is 2020 cm³ correct to 3 s.f.

b)   the capacity of the cylinder $= 2021 \div 1000$ litres

$= 2.021$ litres

$= 2.02$ litres correct to 3 s.f.

In each question from 1 to 8, find the volume of the cylinder whose measurements are given. First make sure that the units are consistent.

1. Radius 6 cm, height 3.8 cm

2. Radius 3.2 cm, height 8 cm

3. Radius 150 cm, height 3 m

4. Radius 0.6 cm, height 9 mm

5. Radius 7.6 cm, height 7.3 cm

6. Radius 28 cm, height 14 cm

7. Radius 58 cm, height 0.07 m

8. Radius 1.2 m, height 68 cm.

9. A cylindrical water tank is of radius 36 cm. How much water is there in the tank when the depth of water is 15 cm? Give your answer in a) cm³ b) litres.

10. A solid cylinder of gold has a radius of 2.5 cm and a height of 2.2 cm. One cubic centimetre of gold has a mass of 19.3 g.

Find a) the volume of the cylinder

b) the mass of the gold.

11. Cement is used to fill cylindrical holes of diameter 20 cm and depth 32 cm.

a) Find the volume of one hole.

b) If the amount of cement available is 201 100 cm³, how many holes will it fill?

Find the radius of a cylinder of volume 72 cm³ and height 9 cm.

$$V = 72 \qquad h = 9$$

$$V = \pi r^2 h$$

$$72 = \pi \times r^2 \times 9$$

$$\frac{72}{\pi \times 9} = r^2$$

$$r^2 = 2.546$$

$$r = 1.595$$

The radius is 1.60 cm correct to 3 s.f.

$$\left(\text{Check: } r^2 \approx \frac{72}{3 \times 9} = \frac{24}{9} \Rightarrow r \approx \frac{5}{3} = 1.\dot{6}\right)$$

In each question from 12 to 17, find the missing measurement of the cylinder.

	Radius	Height	Volume
12.	3.1 cm		72 cm³
13.	11 cm		1024 cm³
14.		1.6 m	15 m³
15.		0.7 m	9.83 m³
16.	3.8 cm		760 cm³
17.		0.12 cm	0.56 cm³

50 litres of water are poured into a cylindrical tank of radius 0.3 m. Find the depth of water in the tank in centimetres.

$$\text{Volume} = 50 \text{ litres}$$
$$= 50\,000 \text{ cm}^3$$
$$V = 50\,000, \ r = 0.3 \times 100 = 30$$
$$V = \pi r^2 h$$
$$50\,000 = \pi \times 30^2 \times h$$
$$\frac{50\,000}{\pi \times 30^2} = h$$
$$h = 17.68$$

The depth of water is 17.7 cm correct to 3 s.f.

18. 1 m³ of water fills a cylindrical drum of radius 50 cm. Find the height of the drum.

19. Water from a full rectangular tank measuring 1 m by 2 m by 0.5 m is emptied into a cylindrical tank and fills it to a depth of 1.2 m.

Find   a) the volume of water involved

b) the diameter of the cylindrical tank.

20. A cylindrical metal rod of radius 1 cm and length 80 cm is melted down and recast into a cylindrical rod of radius 2 cm. How long is the new rod?

21. A cylindrical water butt has a diameter of 80 cm and a height of 1 m. It is half full of water. If a further 20 100 cm³ of water are poured in, find the new depth of water.

22. Water pours out of a cylindrical pipe at the rate of 1 m/s. The diameter of the pipe is 3 cm. How much water comes out in 1 minute?

CONES

VOLUME OF A CONE

We already know that the volume of a pyramid is given by

$$\tfrac{1}{3} \times \text{area of base} \times \text{perpendicular height}$$

where a pyramid is a solid with a flat base and which comes up to a point called the vertex.

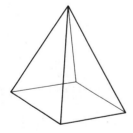 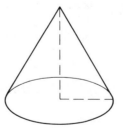 

This definition applies to a cone so the volume of a cone is given by

$$V = \tfrac{1}{3} \times \text{area of circular base} \times \text{perpendicular height}$$

i.e.

$$V = \tfrac{1}{3}\pi r^2 h$$

A cone whose vertex is directly above the centre of the base is called a *right* circular cone; this is the only type of cone that we deal with in this book.

EXERCISE 7g

Find the volume of a cone of base radius 3.2 cm and of height 7.2 cm.

$$r = 3.2 \qquad h = 7.2$$
$$V = \tfrac{1}{3}\pi r^2 h$$
$$= \frac{\pi \times (3.2)^2 \times 7.2}{3}$$
$$= 77.20$$

(Check: $V \approx \dfrac{3 \times 10 \times 7}{3} = 70$)

The volume is 77.2 cm³ correct to 3 s.f.

In each question from 1 to 6 find the volume of the cone whose dimensions are given.

1. Base radius 9 cm, height 20 cm

2. Base radius 2.2 cm, height 5.8 cm

3. Base radius 26.8 cm, height 104 cm

4. Base radius 0.6 cm, height 1.4 cm

5. Base diameter 4.2 cm, height 5.9 cm

6. Base diameter 0.62 m and height 106 cm. Give the volume in cubic metres.

7.

A tower of a toy fort is formed by placing a cone on top of a cylinder. The total height of the tower is 20 cm, the common radius is 5 cm and the height of the cone is 8 cm. Find the volume of the tower.

8.

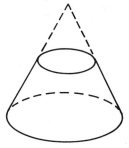

A *frustum* of a cone is formed by cutting the top off a cone.

The original cone has base radius 6 cm and height 10 cm. The part cut off has base radius 3 cm and height 5 cm. Find the volume of the frustum.

9.

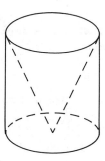

A cylindrical piece of wood of radius 3.6 cm and height 8.4 cm has a conical hole cut in it. The cone has the same radius and the same height as the cylinder. Find the volume of the remaining solid.

10.

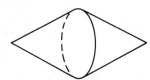

A solid is formed of two equal cones whose diameters are equal to their heights. The distance from the vertex of one cone to the vertex of the other is 12 cm. Find the volume of the solid.

SURFACE AREA OF A CONE

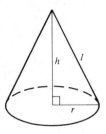

The curved surface area of a cone is given by

$$A = \pi r l$$

where l is the *slant* height.

EXERCISE 7h

In each question from 1 to 4 find the area of the curved surface of the cone whose measurements are given.

1. Radius 4 cm, slant height 10 cm.

2. Radius 9.2 cm, slant height 15 cm

3. Radius 0.6 m, slant height 2.2 m

4. Radius 67 mm, slant height 72 mm

5. Find the total surface area of a cone of base radius 4 cm and with slant height 9 cm.

6. The radius of a cone is 6 cm and its perpendicular height is 8 cm. Find
a) the volume of the cone
b) its slant height
c) its curved surface area.

SPHERES

VOLUME OF A SPHERE

The volume of a sphere is given by the formula

$$V = \tfrac{4}{3}\pi r^3$$

EXERCISE 7i

In each question from 1 to 6 find the volume of the sphere whose radius is given.

1. 3 cm   **3.** 38 cm   **5.** 1.8 m

2. 7.2 cm   **4.** 0.62 cm   **6.** 13 mm

7. Find the volume of a hemisphere of radius 5 cm.

8. Twenty lead spheres of radius 1.2 cm are melted down and recast into a cuboid of length 8 cm and width 4 cm.
a) Find the volume of lead involved.
b) How high is the cuboid?

9. Find, in terms of π, the volume of a sphere of radius $1\frac{1}{2}$ cm. (Do not use a calculator.)

SURFACE AREA OF A SPHERE

The surface area, A, of a sphere of radius r is given by the formula

$$A = 4\pi r^2$$

EXERCISE 7j

In each question from 1 to 4 find the surface area of the sphere whose radius is given.

1. 9 cm   **3.** 41 cm

2. 4.5 cm   **4.** 0.9 cm

5. Find the curved surface area of a hemisphere of radius 23 cm.

6. 240 spheres of radius 0.22 m are to be painted. Each pot of paint contains enough to cover 26 m². How many pots of paint are needed?

PROBLEMS ON VOLUMES OF CONES AND SPHERES ━━━━━━━━━

EXERCISE 7k

1. The radius of a ball-bearing is 0.2 cm. How many ball-bearings can be made from 20 cm³ of metal ?

2.

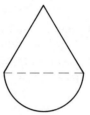

A toy is formed from a cone and a hemisphere. The radius of the hemisphere is 5.2 cm and the total height of the toy is 15 cm. Find the total volume.

3. A hollow metal sphere has an outer radius of 16 cm and its walls are 1 cm thick. Find
a) the inner radius
b) the volume of metal.

4.

A concrete bollard is in the shape of a cylinder surmounted by a hemisphere. The radius of the hemisphere and of the cylinder is 25 cm and the total height is 130 cm. Find the volume of the bollard.

5. Which has the greater volume, a cone of radius 3.5 cm and a perpendicular height of 12 cm or a sphere of radius 3.5 cm ? What is the difference in volume ?

6.

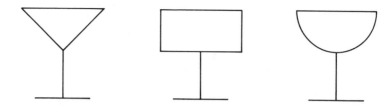

Three glasses are in the shape of a cone, a cylinder and a hemisphere respectively. The radius of each is 4 cm and the depth of the cone and of the cylinder is also 4 cm.

Find, in terms of π, the capacity of
  a) the cone shaped glass
  b) the cylindrical glass
  c) the hemispherical glass.

7. Find in terms of π, the volume of

a) a sphere of radius 2 cm

b) a sphere of radius 8 cm.

The larger sphere is made of metal. It is melted down and made into spheres of radius 2 cm.

c) How many of the smaller spheres can be made from the larger sphere?

PROBLEMS ON SURFACE AREAS OF CONES AND SPHERES

EXERCISE 7I

1. Which has the greater surface area, a cone of radius 3.5 cm and slant height 9 cm or a sphere of radius 3.5 cm? What is the difference between the areas?

2. Find the total surface area of a hemisphere of radius 7 cm.
($\frac{22}{7}$ may be used for π.)

3.

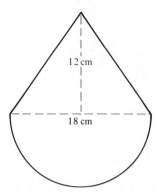

A solid is formed from a cone joined to a hemisphere as shown in the diagram. Find

a) the slant height of the cone

b) the total surface area of the solid.

4. A sphere of radius 1.2 m and a cone of radius 1.2 m and slant height 2.6 m are being painted for a funfair. The tin of paint available contains enough paint to cover 30 m². Is there enough paint for the purpose? Give details of the extra amount needed or the amount of paint left over.

5.

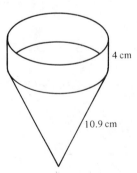

A container is made of sheet metal in the form of an open cone joined to an open-ended cylinder. The radius of the cylinder and of the base of the cone is 8.6 cm, the depth of the cylinder is 4 cm and the slant height of the cone is 10.9 cm. Find the area of sheet metal used.

MIXED EXERCISE

EXERCISE 7m

In each question several alternative answers are given. Write down the letter that corresponds to the correct answer. Do not use your calculator, but remember that using $\pi \approx 3$ gives a quick estimate.

1. The circumference of a circle of radius 10 cm is

    A 18.8 cm   **B** 31.4 cm   **C** 62.8 cm   **D** 314 cm

2. The circumference of a circle is 15 cm. Its diameter is

    A 50.3 cm   **B** 4.77 cm   **C** 2.38 cm   **D** 9.54 cm

3. The area of a circle is 60 cm². Its radius is

    A 4.37 cm   **B** 9.55 cm   **C** 19.1 cm   **D** 3.09 cm

4. A sector of a circle of radius 8 cm contains an angle of 45°. Its area is

    A 25.1 m²   **B** 12.6 cm²   **C** 50.3 cm²   **D** 25.1 cm²

5. An arc of a circle of radius 4 cm subtends an angle of 36° at the centre of the circle. The length of the arc is

    A 50.3 cm   **B** 1.25 cm   **C** 5.03 cm   **D** 2.51 cm

6. A cylinder has radius 2 cm and height 5 cm. The area of its curved surface is

    A 62.8 cm²   **B** 31.4 cm²   **C** 126 cm²   **D** 98.7 cm²

7. A cylinder has radius 2 cm and height 9 cm. Its volume is

    A 509 cm³   **B** 113 cm³   **C** 56.5 cm³   **D** 226 cm³

8. A cylinder has radius 10 cm and height 10 cm. The area of its curved surface is

    A 100π cm³   **B** 1000π cm³   **C** 400π cm³   **D** 200π cm³

9. 400 cm³ of water fills a cylindrical container of radius 4 cm. The height of the container is

    A 63.7 cm   **B** 31.8 cm   **C** 7.96 cm   **D** 15.9 cm

10. A cylinder has radius 3 cm and height 5 cm. Its volume is

    A 30π cm³   **B** 45π cm³   **C** 30π cm²   **D** 48π cm³

8 SIMILAR SHAPES

SIMILAR FIGURES

Similar figures have exactly the same shape, i.e. one figure is an enlargement of the other.

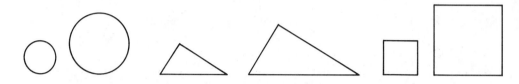

Enlarging a figure does not alter the angles. It does change the lengths of lines, but all the lengths change in the same ratio, i.e. if one line is doubled in length, *all* lines are doubled in length. These facts are particularly useful when we are dealing with similar triangles.

SIMILAR TRIANGLES

When we have a pair of similar triangles, corresponding angles are equal and corresponding sides are in the same ratio.

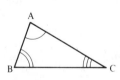 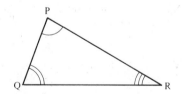

i.e.
$$\widehat{A} = \widehat{P}, \quad \widehat{B} = \widehat{Q}, \quad \widehat{C} = \widehat{R}$$

and
$$\frac{AB}{PQ} = \frac{BC}{QR} = \frac{AC}{PR}$$

$$\left(\text{or } \frac{PQ}{AB} = \frac{QR}{BC} = \frac{PR}{AC} \right)$$

134

It follows that any other pair of corresponding lines has the same ratio as the corresponding sides,

e.g.

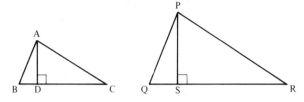

$$\frac{AD}{PS} = \frac{AB}{PQ} \quad \text{and} \quad \frac{\text{perimeter of } \triangle ABC}{\text{perimeter of } \triangle PQR} = \frac{AB}{PQ}$$

To check that two triangles are similar, we need to show that

either   a) the angles of one triangle are equal to the angles of the other triangle

 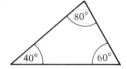

or   b) the three pairs of corresponding sides are in the same ratio

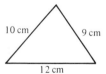

or   c) there is one pair of equal angles and the sides containing these equal angles are in the same ratio.

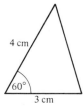

EXERCISE 8a

In triangles PQR and XYZ, $P\widehat{Q}R = 16°$, $R\widehat{P}Q = 42°$, $X\widehat{Y}Z = 16°$ and $X\widehat{Z}Y = 122°$. State whether these triangles are similar.

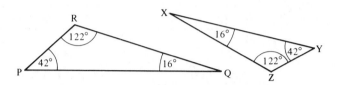

In $\triangle PQR$ $\widehat{R} = 122°$ (angles of a triangle)

In $\triangle XYZ$ $\widehat{Y} = 42°$ (angles of a triangle)

$\therefore$ $\widehat{Q} = \widehat{X}$, $\widehat{P} = \widehat{Y}$ and $\widehat{R} = \widehat{Z}$

$\therefore$ $\triangle$s $\dfrac{PQR}{XYZ}$ are similar

(We line up the corresponding vertices.)

In questions 1 to 8 state whether the two triangles are similar, giving brief reasons:

1.

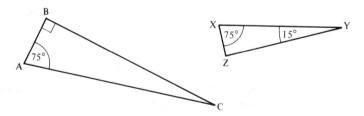

2.

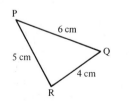
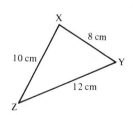

3.

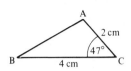

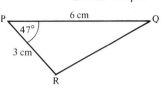

4.

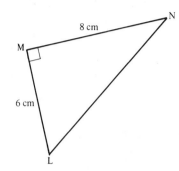

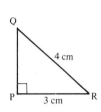

5.

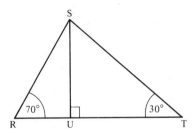

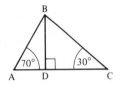

6.

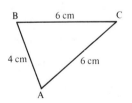

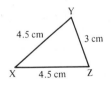

7.

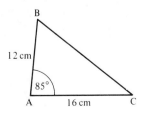

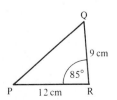

8.

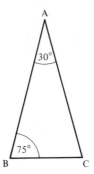

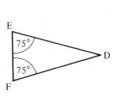

9.

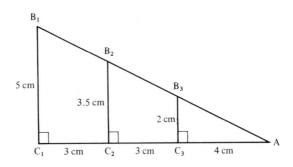

a) Show that $\triangle AB_1C_1$, $\triangle AB_2C_2$ and $\triangle AB_3C_3$ are all similar.

b) Find the values of $\dfrac{B_1C_1}{AC_1}$, $\dfrac{B_2C_2}{AC_2}$ and $\dfrac{B_3C_3}{AC_3}$

10.

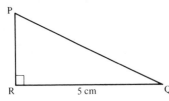

$\widehat{Q}$ is equal to $\widehat{A}$ in question 9.
Is $\triangle PQR$ similar to the triangles in question 9 ?

If it is, write down the value of $\dfrac{PR}{QR}$

and find PR.

11.

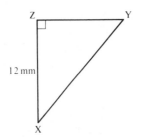

$\widehat{X}$ is equal to $\widehat{A}$ in question 9.

Write down the value of $\dfrac{ZY}{ZX}$ and find ZY.

ABE and DCE are two triangles drawn such that BEC and AED are straight lines and AB is parallel to CD. Show that △ABE is similar to △DEC.

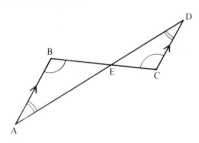

$$\hat{B} = \hat{C} \qquad (\text{alt. } \angle s)$$

$$\hat{A} = \hat{D} \qquad (\text{alt. } \angle s)$$

$$B\hat{E}A = D\hat{E}C \qquad (\text{vert. opp. } \angle s)$$

$$\therefore \ \triangle s \ \frac{BEA}{CED} \ \text{are similar.}$$

12.

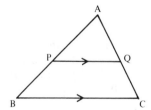

If PQ is parallel to BC, show that △APQ is similar to △ABC.

13.

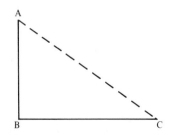 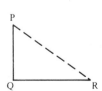

BC is the shadow cast by a flagpole AB. QR is the shadow cast by a stick PQ. Are △ABC and △PQR similar?

14.

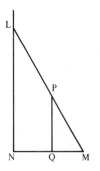

LM is a ladder leaning against a vertical wall with its foot on level ground. PQ is a vertical stick placed so that one end Q is on the ground and the other end P is on the ladder. Show that △LMN and △PMQ are similar.

15.

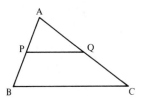

In △ABC, P is the midpoint of AB and Q is the midpoint of AC. Show that △APQ and △ABC are similar.

16.

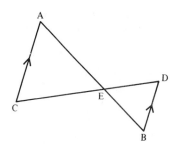

AB and CD are two straight lines that intersect at E in such a way that AC is parallel to DB. Show that △ACE is similar to △BDE.

17.

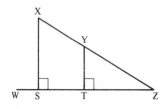

XS and YT are both perpendicular to WZ. Show that △XSZ and △YTZ are similar.

USING SIMILAR TRIANGLES

If we can find a pair of similar triangles we can then use their properties to find angles or lengths of sides.

EXERCISE 8b

In △ABC, X is a point on AB and Y is a point on AC such that AX = 3 cm, XB = 2 cm, AY = 4.5 cm and YC = 3 cm. Show that XY is parallel to BC.

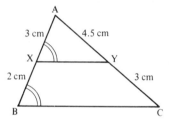

(We will first show that △s $\frac{AXY}{ABC}$ are similar.)

AB = 5 cm and AX = 3 cm

∴ $\frac{AX}{AB} = \frac{3}{5}$

AC = 7.5 cm and AY = 4.5 cm

∴ $\frac{AY}{AC} = \frac{4.5}{7.5} = \frac{45}{75} = \frac{3}{5}$

$\widehat{A}$ is common to △AXY and △ABC

∴   △s $\frac{AXY}{ABC}$ are similar

∴ $A\widehat{X}Y = A\widehat{B}C$

With respect to XY and BC these are corresponding angles.

∴   XY is parallel to BC.

1.

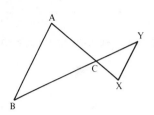

The straight lines AX and BY intersect at C. AC = 4 cm, BC = 6 cm, CY = 3 cm and CX = 2 cm. Show that △ACB and △XCY are similar and hence prove that AB is parallel to XY.

2.

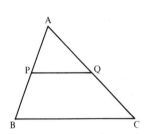

In △ABC, AB = 6 cm and AC = 8 cm. P is the midpoint of AB and Q is the midpoint of AC. Show that △APQ and △ABC are similar. Hence prove that PQ is parallel to BC.

3.

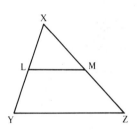

In △XYZ, L is the midpoint of XY and M is the midpoint of XZ. Write down the values of $\dfrac{XL}{XY}$ and $\dfrac{XM}{XZ}$. Are △XLM and △XYZ similar? Is LM parallel to YZ?

4.

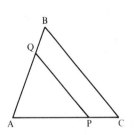

In △ABC, P is a point on AC and Q is a point on AB such that
$$\frac{AQ}{AB} = \frac{2}{3} \quad \text{and} \quad \frac{AP}{AC} = \frac{2}{3}$$
Show that △APQ is similar to △ACB.
If $\widehat{A} = 70°$ and $\widehat{APQ} = 50°$, find $\widehat{ABC}$.

5.

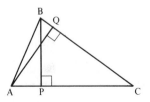

In △ABC, AQ is perpendicular to BC and BP is perpendicular to AC. Show that △BPC and △AQC are similar. Hence show that
$$\frac{CP}{CQ} = \frac{BC}{AC}$$

6. Use the diagram and the results from question 5 to show that, if PQ is joined, △CPQ is similar to △CBA. Is PQ parallel to AB?

In △ABC, P is a point on AB and Q is a point on AC such that PQ is parallel to BC. AP = 5 cm, PB = 2 cm and AQ = 7 cm. Find AC.

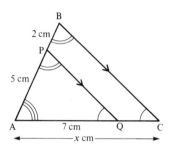

$$A\widehat{B}C = A\widehat{P}Q \quad (\text{corresponding} \angle s)$$

$$A\widehat{C}B = A\widehat{Q}P \quad (\text{corresponding} \angle s)$$

$\widehat{A}$ is common to △ABC and △APQ

∴   △s $\begin{matrix} ABC \\ APQ \end{matrix}$ are similar.

∴ $\dfrac{AC}{AQ} = \dfrac{AB}{AP}$

∴ $\dfrac{x}{7} = \dfrac{7}{5}$

$$x = \dfrac{7}{5} \times 7$$

$$= \dfrac{49}{5}$$

$$= 9.8$$

∴ AC = 9.8 cm

7.

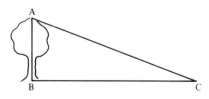

The shadow cast by a tree AB is BC, where BC = 20 m. The shadow cast by a stick PQ is QR, where QR = 4 m. If PQ = 1 m, find the height of the tree.

8. Try using the method described in question 7 to find the height of a tree, lamp post or building near you. You will need a one metre rule, a long tape measure (or use the rule and a piece of chalk) and a sunny day!

9.

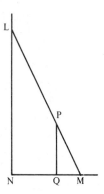

LM is a ladder leaning against a vertical wall LN with its foot, M, on level ground such that NM = 1.5 m. PQ is a straight stick placed so that it is vertical, with one end P on the ladder and the other end Q resting on the ground. If QM = 0.5 m and PM = 2 m, find the length of the ladder.

10. Here is a way to find (roughly) the width of a river without crossing it. Place a stake B on one bank opposite a landmark A (such as a tree) on the other bank. Walk a distance of 5 m along the bank at right angles to AB and place another stake C. Walk another 1 m in the same direction and place another stake D. Now walk at right angles to BD until A and C are in line, then place a stake E. The width of the river is five times the distance between D and E. Draw a diagram showing this and explain why it works.

11. In △ABC, P is a point on AC and Q is a point on BC such that PQ is parallel to AB.
a) Show that △ABC and △PQC are similar.
b) If AC = 8 cm, BC = 6 cm and PC = 5 cm find QC.
c) Find BQ.
d) Write down the values of AP : PC and BQ : QC.

12.

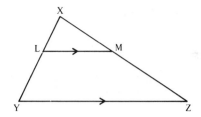

L and M are points on the sides XY and XZ respectvely of a triangle XYZ such that LM is parallel to YZ.

a) Show that △XLM and △XYZ are similar.

b) If XY = 6 cm, XZ = 10 cm and XM = 4 cm, find XL.

c) Find MZ.

d) Write down the values of XL : LY and XM : MZ.

13.

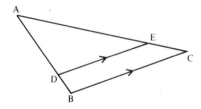

In △ABC, D is a point on AB and E is a point on AC such that DE is parallel to BC.

a) Show that △ADE and △ABC are similar.

b) If AE = 8 cm, AC = 10 cm and AB = 5 cm, find AD.

c) Write down the values of AD : DB and AE : EC.

14. In △ABC, X is a point on AB and Y is a point on AC such that XY is parallel to BC.

a) Show that △AXY and △ABC are similar.

b) If AB = 24 cm, AC = 30 cm and AX = 8 cm, find AY.

c) Write down the values of AX : XB and AY : YC.

THE INTERCEPT THEOREM

From the last exercise we can see that

> if a line is drawn parallel to one side of a triangle it divides the other two sides in the same ratio.

This is known as the intercept theorem, and can now be used.

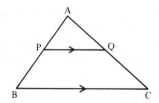

For example, in $\triangle ABC$, PQ is parallel to BC.

$\therefore$ $$\frac{AP}{PB} = \frac{AQ}{QC} \quad \text{(intercept theorem)}$$

EXERCISE 8c

In $\triangle LMN$, XY is drawn parallel to MN such that $LX = 8\,cm$, $LY = 12\,cm$ and $XM = 3\,cm$. Find YN.

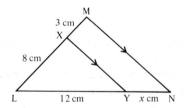

Let YN be x cm

$$\frac{x}{12} = \frac{3}{8} \quad \text{(intercept theorem)}$$

$${}^{1}\!\!\not{12} \times \frac{x}{\not{12}_{1}} = {}^{1}\!\!\not{12} \times \frac{3}{\not{8}_{2}}$$

$$x = 4.5$$

$\therefore$ $$YN = 4.5\,cm$$

In each question from 1 to 8, find the length of the line marked *x*. All measurements are in centimetres.

1.

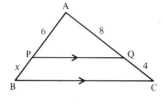

2.

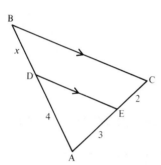

3.

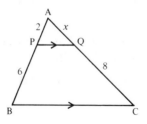

4.

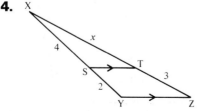

5.

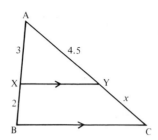

6.

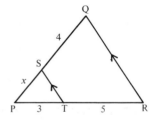

7.

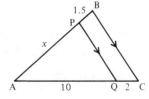

8.

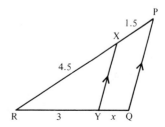

9.

In △ABC, PQ is drawn parallel to AB such that AC = 18 cm and PC = 6 cm. If BC = 12 cm, find BQ.

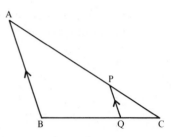

10.

In △XYZ, PQ is drawn parallel to YZ. If XQ = 12 cm, QZ = 3 cm and XY = 10 cm, find PY.

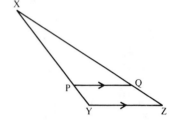

DIVIDING A LINE INTO A GIVEN NUMBER OF EQUAL PARTS

Suppose that we want to divide the line drawn below into three equal parts.

We can do it with a ruler, i.e. measure AB, calculate one third of its length and then measure this distance from A and from B.

Using the intercept theorem, however, we can divide AB into three equal parts without any measuring.

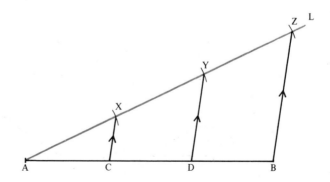

1. Draw a line AL at an angle to AB. (Any angle will do but one of about 30° is convenient.)

2. Open your compasses to any radius. (Roughly 3 cm is a convenient radius to work with in this case.)

3. With the point on A draw an arc to cut AL at X.

4. Without changing the radius, move the point to X and draw an arc to cut AL at Y.

5. Without changing the radius, move the point to Y and draw an arc to cut AL at Z.

6. Join ZB.

7. Draw lines parallel to ZB through X and Y to cut AB at C and D.

Now C and D divide AB into three equal lengths.

To divide AB into five equal parts, we cut five equal lengths off AL.

EXERCISE 8d

1. Describe some advantages and disadvantages of the two methods given for dividing a line into a number of equal parts.

2. Draw a line 10 cm long. Divide it into three equal parts by
a) calculation and measurement   b) construction.

3. Draw a line 9 cm long. Divide it into five equal parts by
a) calculation and measurement   b) construction.
Which of these two methods looks more accurate?

4. Draw a line AB that is 11 cm long. Divide AB into five equal parts, by construction.

5. In the diagram, P, Q, R and S divide the line AB into five equal parts.

a) If the length of AP is x cm, write down, in terms of x, the lengths of AQ, AR, AS and AB.

b) Write down the values of the ratios AP : PB,   AQ : QB,   AR : RB, AS : SB.

6. Draw a line AB that is 12 cm long.
a) Divide AB into three equal parts by calculation and measurement.
b) Mark the point P on AB such that AP : PB = 2 : 1.
(The point P is said to divide AB in the ratio 2 : 1.)

7. Draw a line AB that is 15 cm long.

a) Divide AB into five equal parts by calculation and measurement.

b) Mark the point P on AB such that $AP:PB = 2:3$.
   (The point P is said to divide AB in the ratio $2:3$.)

8. Draw a line AB that is 13 cm long.

a) Divide AB into three equal parts by construction.

b) Mark the point P on AB such that $AP:PB = 1:2$.

c) Is it necessary to draw all three parallel lines in order to find P?

9.

P is a point on the line XY such that $XP:PY = 3:4$. To find the point P, into how many equal parts must the line XY be divided?

a) Draw a line XY that is 14 cm long. Find P by measurement.

b) Draw a line XY that is 16 cm long. Find P by construction, but draw only those parallel lines that are necessary.

AREAS OF SIMILAR FIGURES

These four squares are similar.

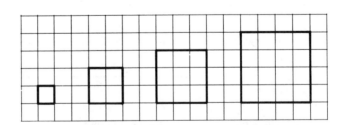

The ratio of the lengths of their sides is $1:2:3:4$. Counting squares gives the ratio of their areas as $1:4:9:16$.

But $1:4:9:16 = 1^2:2^2:3^2:4^2$

i.e. the ratio of the areas is equal to the ratio of the squares of corresponding lengths.

EXERCISE 8e

1. These four rectangles are similar.

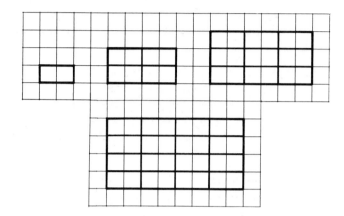

a) Write down the ratio of the lengths of their bases.

b) By counting rectangles, write down the ratio of their areas.

Is there a relationship between these two ratios?

2. These four parallelograms are similar.

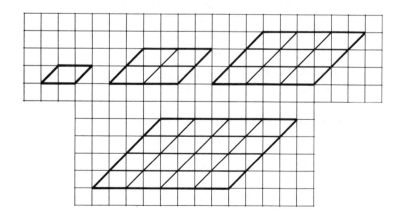

a) Write down the ratio of the lengths of their bases.

b) By counting parallelograms, write down the ratio of their areas.

Is there a relationship between these two ratios?

3. These four triangles are similar.

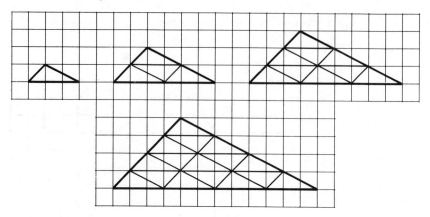

a) Write down the ratio of the lengths of their bases.

b) By counting triangles, write down the ratio of their areas.

What is the relationship between these two ratios?

4.

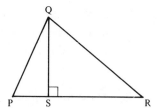

Triangles ABC and PQR are similar and the scale factor for enlarging △ABC to △PQR is 2, i.e. AB : PQ = 1 : 2.

a) If AC = 10 cm, find PR.

b) If BD = 6 cm, find QS.

c) Find the area of each triangle.

d) Find the ratio of the areas of these triangles.

In questions 5 to 8, the pairs of figures are similar. First find the length required and then find the area of each figure.

5. Find XW.

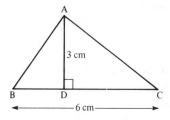
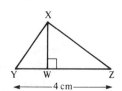

6. Find LN.

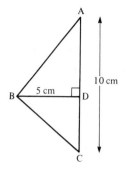

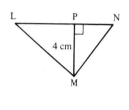

7. Find BC.

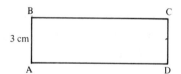

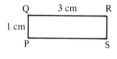

8. Find LP.

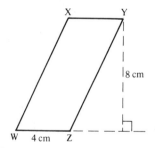

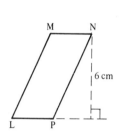

9. Using your answers to questions 5 to 8, complete the following table.

Similar figures	Ratio of sides	Ratio of areas
Triangles in question 5		
Triangles in question 6		
Rectangles in question 7		
Parallelograms in question 8		

What is the relationship between the ratio of the areas and the ratio of corresponding sides for each of these pairs of similar figures ?

THE RELATIONSHIP BETWEEN THE AREAS OF SIMILAR FIGURES

> For similar figures, the ratio of their areas is equal to the ratio of the squares of corresponding lengths.

EXERCISE 8f

Parallelograms ABCD and PQRS are similar.
If CD = 8 cm and RS = 12 cm, what is the value of the ratio of area ABCD to area PQRS?

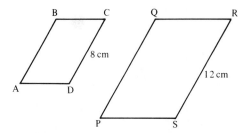

CD and RS are corresponding sides and

$$\frac{CD}{RS} = \frac{\overset{2}{\cancel{8}}}{\underset{3}{\cancel{12}}}$$

$$= \frac{2}{3}$$

∴

$$\frac{\text{area ABCD}}{\text{area PQRS}} = \frac{2^2}{3^2}$$

$$= \frac{4}{9}$$

For each pair of similar figures in questions 1 to 6, write down the ratio of their areas:

1.

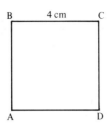

2.

 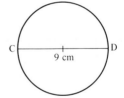

3.

 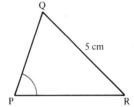

4.

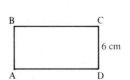 

5.

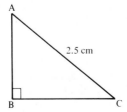 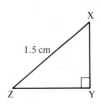

6.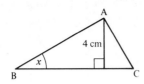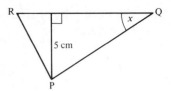

In questions 7 to 9 the pictures are from catalogues, but the dimensions given are those of the actual object.

7.

700 mm

The width of the picture of the door is 20 mm. Find the ratio of the area of the picture of the door to the area of the actual door.

8.

60 cm

The width of the drawing of the chest is 3 cm. Find the ratio of the area of the front of the chest in the picture to the area of the front of the actual chest of drawers.

9.

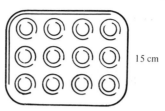

15 cm

The dimension marked 15 cm in the picture of the tin is $2\frac{1}{2}$ cm. Find the ratio of the area of the catalogue picture to the area of the real tin.

Triangles ABC and XYZ are similar and $\widehat{C} = \widehat{Z}$. If the area of △ABC is 3.2 cm² and the area of △XYZ is 1.8 cm², find the value of AB : XY.

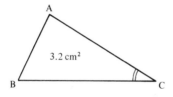

A

3.2 cm²

B   C

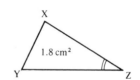

X

1.8 cm²

Y   Z

$$\frac{\text{area ABC}}{\text{area XYZ}} = \frac{3.2}{1.8}$$

$$= \frac{\cancel{32}^{16}}{\cancel{18}_{9}}$$

$$= \frac{16}{9}$$

but

$$\frac{\text{area ABC}}{\text{area XYZ}} = \frac{AB^2}{XY^2}$$

$\therefore$

$$\frac{AB^2}{XY^2} = \frac{16}{9}$$

$\therefore$

$$\frac{AB}{XY} = \frac{4}{3}$$

Find the value of **AB : XY** for each of the following pairs of similar figures.

10.

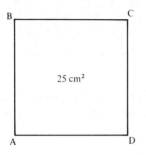

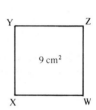

11.

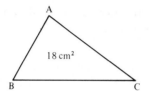

12.

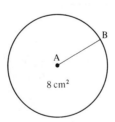

13.

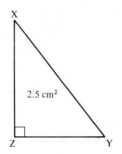

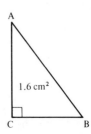

14.

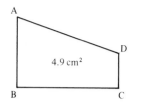

15.

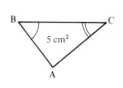
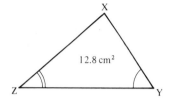

Triangles ABC and PQR are similar, with $\widehat{A} = \widehat{P}$ and $\widehat{C} = \widehat{R}$. If AC = 4 cm, PR = 3 cm and area $\triangle PQR = 4.5\,cm^2$, find area $\triangle ABC$.

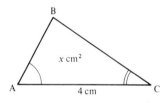
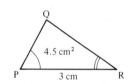

AC and PR are corresponding sides and $\dfrac{AC}{PR} = \dfrac{4}{3}$

$\therefore$ $\dfrac{\text{area } \triangle ABC}{\text{area } \triangle PQR} = \dfrac{16}{9}$

i.e. $\dfrac{x}{4.5} = \dfrac{16}{9}$

i.e. $4.5 \times \dfrac{x}{4.5} = \dfrac{16}{9} \times 4.5$

$x = 8$

$\therefore$ area $\triangle ABC = 8\,cm^2$

16.

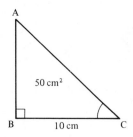

Triangles ABC and XYZ are similar. From the information given in the diagram, find the area of △XYZ.

17.

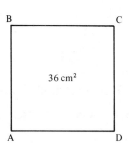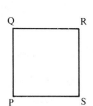

ABCD and PQRS are squares and AB : PQ = 3 : 2. If the area of ABCD is 36 cm², find the area of PQRS.

18. Rectangles ABCD and WXYZ are similar. If BD = 4 cm, XZ = 5 cm and area ABCD = 4.8 cm², find the area of WXYZ.

19.

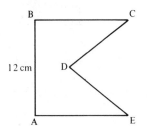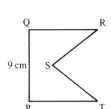

ABCDE and PQRST are similar shapes. If AB = 12 cm, PQ = 9 cm and area PQRST = 36 cm², find the area of ABCDE.

20. Triangles ABC and XYZ are both equilateral triangles.
If area $\triangle$ABC : area $\triangle$XYZ = 36 : 25, find the value of AB : XY.

21. Rectangles ABCD and PQRS are similar. If area ABCD = $2.5\,\text{cm}^2$,
area PQRS = $4.9\,\text{cm}^2$ and AB = $1.5\,\text{cm}$, find PQ.

22.

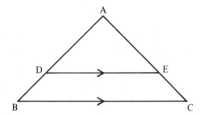

DE is parallel to BC. If AE = $10\,\text{cm}$, EC = $4\,\text{cm}$ and the area of $\triangle$ABC
is $98\,\text{cm}^2$, find the area of $\triangle$ADE.

HARDER EXAMPLES

EXERCISE 8g

In this exercise, it may be necessary first to show that the triangles are
similar.

1.

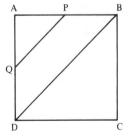

ABCD is a square of area $12\,\text{cm}^2$. P is the
midpoint of AB and Q is the midpoint
of AD. Find the area of $\triangle$APQ.

2. Triangles ABC and PQR are similar. If the area of $\triangle$ABC is four times that
of $\triangle$PQR and AB and PQ are corresponding sides, what is the value of
AB : PQ ?

3. The scale of a map is 1 : 1000. On the map, the area representing a mansion
is $2\,\text{cm}^2$. What is the actual area in m^2 occupied by the mansion ?

4.

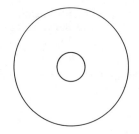

The area of the larger circle is sixteen times that of the smaller circle. What is the ratio of the radii of the two circles?

5. ABC is a triangle with X a point on AB and Y a point on AC such that XY is parallel to BC. If AY = 3 cm, YC = 4 cm and XB = 3 cm, find

a) AX b) $\dfrac{\text{area } \triangle AXY}{\text{area } \triangle ABC}$ c) $\dfrac{\text{area } \triangle AXY}{\text{area trapezium XYCB}}$.

6.

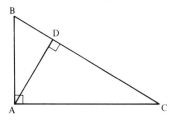

Triangle ABC has a right angle at A and AD is perpendicular to BC. The area of △ABD is 4 cm² and the area of △ADC is 5 cm². Find the ratio of the area of △ABD to the area of △ABC and hence the value of AB : BC.

7.

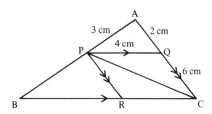

In the diagram, PQ is parallel to BC and PR is parallel to AC.
AQ = 2 cm, QC = 6 cm, AP = 3 cm and PQ = 4 cm

a) Calculate i) PB ii) BR iii) $\dfrac{\text{area } \triangle APQ}{\text{area } \triangle ABC}$ iv) $\dfrac{\text{area } \triangle BPR}{\text{area } \triangle ABC}$

b) If the area of triangle APQ is a cm², express in terms of a
   i) area △ABC ii) area △CPQ

8. Construct a triangle ABC such that BC = 10 cm, AC = 9 cm and AB = 6 cm. Find a point D on AB and a point E on AC, such that DE is parallel to BC and the area of △ADE is one ninth of the area of △ABC.

VOLUMES OF SIMILAR SHAPES

EXERCISE 8g

These four cubes are similar.

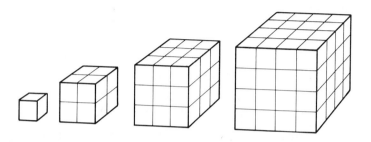

The ratios of the lengths of their sides is $1:2:3:4$. By counting cubes, the ratio of their volumes is $1:8:27:64$

$$\text{But } 1:8:27:64 = 1^3:2^3:3^3:4^3$$

i.e. the ratio of the volumes is equal to the ratio of the cubes of corresponding lengths.

EXERCISE 8h

1. These three solids are similar.

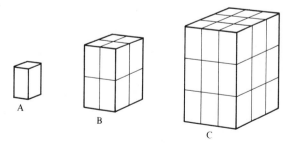

a) Write down the ratio of the lengths of the bases.

b) Write down the ratio of the lengths of the heights.

c) By counting cuboids equal in shape and size to the cuboid given in A, write down the ratio of the volumes.

Is there a relationship between your answers to (a), (b) and (c)?

2.

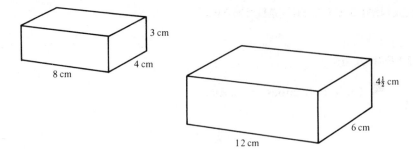

These are two similar rectangular blocks.

a) Write down the ratio of their
   i) longest edges   ii) depths   iii) heights.

b) By counting cubes of side 1 cm write down the ratio of their volumes.

Is there any relationship between the ratios in (a) and (b)?

3.

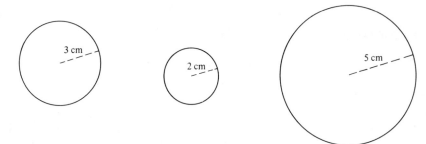

These three solids are spheres.

a) Write down the ratio of the radii of the three spheres.

b) If the volume of a sphere of radius r is given by the formula $V = \frac{4}{3}\pi r^3$
   express the volume of each sphere as a multiple of π. Hence write down
   the ratio of their volumes.

Is there a relationship between the ratio found in (a) and the ratio found
in (b)?

4. These three cans are similar cylinders.

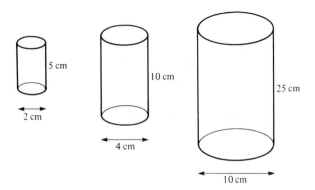

a) Write down the ratio of

i) their heights   ii) their base radii

b) If the volume of a cylinder of height h and base radius r is given by the formula $V = \pi r^2 h$, express the capacity of each can in terms of π. Hence find the ratio of their capacities.

Is there a relationship between these ratios?

THE RELATIONSHIP BETWEEN THE VOLUMES OF SIMILAR SHAPES

> For similar solids, the ratio of their volumes is equal to the ratio of the cubes of corresponding linear dimensions.

In the same way, the ratio of the capacities of similar containers is equal to the ratio of the cubes of corresponding linear dimensions.

In problems where ratios are used to find an unknown quantity it is wise to write down the ratio so that the unknown quantity is in the numerator rather than in the denominator.

EXERCISE 8i

A sculptor is commissioned to create a bronze statue 2 m high. He begins by making a clay model 30 cm high.

a) Express, in its simplest form, the ratio of the height of the completed bronze statue to the height of the clay model.

b) If the total surface area of the model is 360 cm², find the total surface area of the statue.

c) If the volume of the model is 1000 cm³ find the volume of the statue.

a) $\dfrac{\text{height of bronze statue}}{\text{height of clay model}} = \dfrac{200}{30} = \dfrac{20}{3}$

b) $\dfrac{\text{surface area of statue}}{\text{surface area of model}} = \dfrac{20^2}{3^2}$

i.e. $\dfrac{\text{surface area of statue}}{360} = \dfrac{400}{9}$

$\therefore$ surface area of statue $= \dfrac{400}{9}^{1} \times 360^{40}\,\text{cm}^2$

$= 16\,000\,\text{cm}^2$

$= 1.6\,\text{m}^2$

c) $\dfrac{\text{volume of statue}}{\text{volume of model}} = \dfrac{20^3}{3^3}$

i.e. $\dfrac{\text{volume of statue}}{1000} = \dfrac{8000}{27}$

$\therefore$ volume of statue $= \dfrac{8000}{27} \times 1000\,\text{cm}^3$

$= \dfrac{8000 \times 1000}{27 \times 1\,000\,000}\,\text{m}^2$

$= \dfrac{8}{27}\,\text{m}^3$

In this exercise objects referred to in the same question are mathematically similar.

1. The sides of two cubes are in the ratio $2:1$. What is the ratio of their volumes?

2. The radii of two spheres are in the ratio $3:4$. What is the ratio of their volumes?

3. Two regular tetrahedrons have volumes in the ratio $8:27$. What is the ratio of their sides?

4. Two right cones have volumes in the ratio $64:27$. What is the ratio of
a) their heights   b) their base radii?

5. Two similar bottles are such that one is twice as high as the other. What is the ratio of   a) their surface areas   b) their capacities?

6. Each linear dimension of a model car is $\frac{1}{10}$ of the corresponding car dimension. Find the ratio of
a) the areas of their windscreens   b) the capacities of their boots
c) the widths of the cars   d) the number of wheels they have.

7. Three similar jugs have heights $8\,\text{cm}$, $12\,\text{cm}$ and $16\,\text{cm}$. If the smallest jug holds $\frac{1}{2}$ pint, find the capacities of the other two.

8. A cylindrical cola can $10\,\text{cm}$ high costs $2\,\text{p}$ to make. What is the cost of a can standing $15\,\text{cm}$ high?

9. Three similar drinking glasses have heights $7.5\,\text{cm}$, $9\,\text{cm}$ and $10.5\,\text{cm}$. If the tallest glass holds 343 centilitres find the capacities of the other two.

10. The capacities of three similar jugs are $486\,\text{cl}$, $1152\,\text{cl}$ and $2250\,\text{cl}$.
a) If the jug with the largest capacity is $15\,\text{cm}$ high, find the heights of the other two.   b) If the base area of the smallest jug is $36\,\text{cm}^2$ find the base areas of the other two.

11. A toy manufacturer produces model cars which are similar in every way to the actual cars. If the ratio of the door area of the model to the door area of the car is $1:2500$ find
a) the ratio of their lengths
b) the ratio of the capacities of their petrol tanks
c) the width of the model, if the actual car is $150\,\text{cm}$ wide
d) the area of the rear window of the actual car to the area of the rear window of the model is $3\,\text{cm}^2$.

12. The ratio of the areas of two similar labels on two similar jars of coffee is $144:169$. Find the ratio of   a) the heights of the two jars   b) their capacities.

13. A wax model has a mass of $1\,\text{kg}$. Find the mass of a similar model which is twice as tall and made from metal eight times as heavy as wax.

The radius of a spherical soap bubble increases by 5%. Find, correct to the nearest whole number, the percentage increase in
a) its surface area   b) its volume.

If the original radius is r the increased radius is $\dfrac{105}{100} \times r$ i.e. $1.05r$

Therefore $\dfrac{\text{new radius}}{\text{old radius}} = \dfrac{1.05r}{r}$

$= \dfrac{1.05}{1}$

a) $\dfrac{\text{new surface area}}{\text{original surface area}} = \dfrac{(1.05)^2}{1^2}$

$= 1.1025$

i.e. new surface area $= 1.1025$ of the original surface area

i.e. new surface area $= 1.1025 \times 100\%$ of original surface area

$= 110.25\%$ of the original surface area.

The surface area has therefore increased by 10%.

b) $\dfrac{\text{new volume}}{\text{original volume}} = \dfrac{(1.05)^3}{1^3}$

$= \dfrac{1.158}{1}$

$= 1.158$

i.e. new volume $= 1.158$ of the original volume

$= 115.8\%$ of the original volume

The volume has therefore increased by 16%.

14. The radius of one sphere is 10% more than the radius of another. Find, correct to the nearest whole number, the percentage difference in
a) their surface areas   b) their volumes.

15. The radius of a spherical snowball increases by 80%. Find, correct to the nearest whole number, the percentage increase in   a) its surface area   b) its volume.

16. The side of one cube is 20% greater than the side of another. Find, correct to the nearest whole number, the percentage difference in   a) their surface areas   b) their volumes.

17. The volume of a cone increases by 100%. Find the percentage increase in
a) its height   b) its base radius   c) its surface area.

Give your answers correct to the nearest whole number.

18. A spherical grapefruit has a diameter of 10 cm. If its peel is 1 cm thick, find, correct to the nearest whole number, the percentage of the volume of the grapefruit that is thrown away as peel.

MIXED EXERCISES

EXERCISE 8j

1. In △ABC, P is a point on AB and Q is a point on AC such that AP = 5 cm, PB = 10 cm, AQ = 4 cm and QC = 8 cm. Show that PQ is parallel to BC.

2. The ratio of the areas of two similar triangles is 49 : 25. What is the ratio of corresponding sides?

3. In △XYZ, P is a point on XY and Q is a point on XZ such that PQ is parallel to YZ.

a) Show that △XPQ and △XYZ are similar

b) If XY = 36 cm, XZ = 30 cm and XP = 24 cm,
   find   i) XQ   ii) QZ

c) Write down the values of XP : PY and PQ : YZ.

4. Draw a line 9.7 cm long. Use a constructional method to divide it into four equal parts.

5.

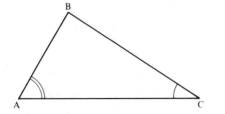
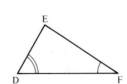

Triangles ABC and DEF are similar. If the area of △ABC is 12.5 cm², the area of △DEF is 4.5 cm², and AB = 5 cm find

a) DE   b) the value of AC : DF   c) the value of EF : BC.

EXERCISE 8k

1. Two similar rectangles have areas in the ratio 49 : 81. What is the ratio of
a) their longer sides   b) their shorter sides ?

2.

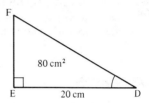

Triangles ABC and DEF are similar. From the information given in the diagrams

a) the area of △ABC   b) the length of FE   c) ratio AC : FD.

3.

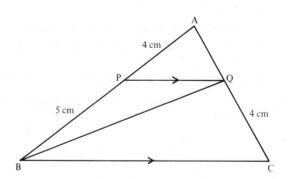

Use the information given in the diagram to find

a) AQ   b) PQ : BC

c) $\dfrac{\text{area } \triangle APQ}{\text{area } \triangle ABC}$ d) $\dfrac{\text{area } \triangle APQ}{\text{trapezium PQCB}}$

e) $\dfrac{\text{area } \triangle APQ}{\text{area } \triangle BPQ}$ f) $\dfrac{\text{area } \triangle BPQ}{\text{area } \triangle BCQ}$

4.

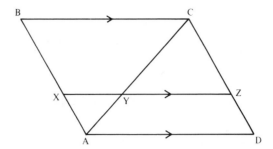

AC is the diagonal of a rhombus ABCD. The line XYZ is parallel to AD, AX = 3 cm and AB = 9 cm. Find

a) $\dfrac{XY}{BC}$ b) $\dfrac{AY}{AC}$ c) $\dfrac{CY}{AC}$ d) $\dfrac{YZ}{AD}$

e) $\dfrac{\text{area } \triangle AXY}{\text{area } \triangle ABC}$ f) $\dfrac{\text{area } \triangle CYZ}{\text{area } \triangle ACD}$

5. An inverted hollow cone is filled with water to half its depth. What fraction of the available capacity is filled?

9 INFORMATION MATRICES

MATRICES AS STORES OF INFORMATION

We have mainly used matrices for describing and performing transformations but another application of matrices is in computer work.

We have looked at one such application: the use of matrices for solving simultaneous equations. The advantage of this method for computer solutions is that it does not require decisions to be made. The fact that it is generally the longest method does not matter when computers do the solution because computers work fast!

Another widespread application of matrices in computer work is to store information.

For example, a hotel supplies three types of packed meal, A, B, and C. On Saturday, 5 meals of type A, 10 meals of type B and 7 meals of type C are ordered. On Sunday, 12 meals of type A, 6 meals of type B and 9 meals of type C are ordered. This information can be displayed in a matrix, **M**, where

$$\mathbf{M} = \begin{matrix} & \begin{matrix} A & B & C \end{matrix} \\ \begin{matrix} \text{Sat} \\ \text{Sun} \end{matrix} & \begin{pmatrix} 5 & 10 & 7 \\ 12 & 6 & \textcircled{9} \end{pmatrix} \end{matrix}$$

The columns represent the different meals and the rows represent the different days. The ringed entry indicates that 9 type C meals were ordered on Sunday.

We chose to use a 2×3 matrix, but we could equally well represent the information in a 3×2 matrix, **N**, where

$$\mathbf{N} = \begin{matrix} & \begin{matrix} \text{Sat} & \text{Sun} \end{matrix} \\ \begin{matrix} A \\ B \\ C \end{matrix} & \begin{pmatrix} 5 & 12 \\ 10 & 6 \\ 7 & 9 \end{pmatrix} \end{matrix}$$

EXERCISE 9a

1. The matrix A shows the number of acres of land used for different purposes on two farms, A and B.

$$\begin{array}{c} & \text{Wheat} & \text{Grazing} & \text{Other crops} \\ \mathbf{A} = \begin{array}{c} \text{Farm A} \\ \text{Farm B} \end{array} \left(\begin{array}{ccc} 100 & 300 & 50 \\ 200 & 0 & 300 \end{array} \right) \end{array}$$

Copy this matrix and

a) ring the entry that gives the number of acres on Farm A used for other crops

b) put a square round the entry that gives the number of acres used for wheat on Farm B

c) find the total number of acres used for growing wheat on both farms.

2. The matrix T shows the number of packets of different brands of coffee sold in one week in three supermarkets A, B and C.

$$\begin{array}{c} & \text{Brand X} & \text{Brand Y} & \text{Brand Z} \\ \mathbf{T} = \begin{array}{c} \text{A} \\ \text{B} \\ \text{C} \end{array} \left(\begin{array}{ccc} 50 & 25 & 37 \\ 100 & 150 & 89 \\ 92 & 250 & 340 \end{array} \right) \end{array}$$

Copy this matrix and

a) ring the entry that gives the number of packets of Brand Y sold in Supermarket C

b) put a square round the entry that gives the number of packets of Brand X sold in Supermarket A

c) find the total number of packets sold in Supermarket B.

3. A carpet manufacturer makes three grades of carpet. The matrix shows the number of metres of each grade ordered in each of four consecutive weeks.

$$\begin{array}{c} & \text{Grade I} & \text{Grade II} & \text{Grade III} \\ \begin{array}{c} \text{Week 1} \\ \text{Week 2} \\ \text{Week 3} \\ \text{Week 4} \end{array} \left(\begin{array}{ccc} 200 & 150 & 120 \\ 350 & 200 & 70 \\ 190 & 250 & 100 \\ 280 & 210 & 110 \end{array} \right) \end{array}$$

Copy this matrix and give its size.

a) Ring the entry that gives the number of metres of Grade II carpet ordered in week 3.

b) Underline the entry that gives the number of metres of Grade III carpet ordered in week 2.

c) How many metres of grade III carpet were ordered over the four week period?

d) How many metres of carpet were ordered in week 2?

e) Show the same information in a 3×4 matrix, and underline the entry giving the number of metres of grade II carpet ordered in week 4.

4. In shop A, potatoes cost 10p per lb, carrots cost 8p per lb and parsnips cost 12p per lb. In shop B, potatoes are 12p per lb, carrots are 9p per lb and parsnips are 10p per lb.

a) Show this information in a 2×3 matrix, **M**.

b) Ring the entry giving the cost of carrots in shop **B**.

c) Mr Smith buys 5lb of potatoes, 1lb of carrots and 2lb of parsnips. Show this information in a column matrix, **P**.

d) How much would Mr Smith's purchases cost in shop **A**?

e) Find the product **MP**.

f) What meaning can you give to the two entries in the product **MP**?

5. A school's supplier stocks chalk in three boxes of different sizes; Box A which contains 20 sticks of chalk, Box B which contains 50 sticks of chalk and Box C which contains 100 sticks of chalk. The matrix **M** shows the numbers of each box supplied in each of three months.

$$\mathbf{M} = \begin{array}{c} \\ A \\ B \\ C \end{array} \begin{array}{ccc} \text{Jan} & \text{Feb} & \text{Mar} \\ \begin{pmatrix} 200 & 100 & 200 \\ 50 & 10 & 20 \\ 150 & 70 & 100 \end{pmatrix} \end{array}$$

a) How many sticks of chalk are supplied in January?

b) How many sticks of chalk are supplied in March?

c) Write down a 1×3 matrix, **N**, showing the number of sticks of chalk in each type of box.

d) Can you find a way of multiplying N and M together to give the information asked for in parts (a) and (b)?

6. A vending machine accepts ten-pence, twenty-pence, and fifty-pence coins only. The matrix **A** shows the numbers of each coin in the machine when emptied on two separate occasions.

$$\mathbf{A} = \begin{array}{c} \\ \text{1st emptying} \\ \text{2nd emptying} \end{array} \begin{array}{ccc} 10p & 20p & 50p \\ \begin{pmatrix} 50 & 10 & 20 \\ 70 & 30 & 80 \end{pmatrix} \end{array}$$

a) How many coins were in the machine the first time it was emptied?

b) Evaluate $\mathbf{A} \begin{pmatrix} 1 \\ 1 \\ 1 \end{pmatrix}$ and give a meaning to the entries.

c) How much money was in the machine the first time it was emptied?

d) Write down a column matrix, **V**, giving the value of each coin.

e) Can you multiply **A** and **V** together so that the entries in the result give the amount of money in the machine on each occasion that it was emptied?

GETTING INFORMATION FROM MATRICES

Consider again the matrix **M**, showing the numbers of each type of packed meal ordered on each day of a weekend where

$$\mathbf{M} = \begin{array}{c} \\ \text{Sat} \\ \text{Sun} \end{array} \begin{pmatrix} 5 & 10 & 7 \\ 12 & 6 & 9 \end{pmatrix} \begin{array}{c} \text{A} \quad \text{B} \quad \text{C} \\ \\ \end{array}$$

The matrix **P** gives the cost of each type of meal, where

$$\mathbf{P} = \begin{array}{c} \text{A} \\ \text{B} \\ \text{C} \end{array} \begin{pmatrix} 5 \\ 3 \\ 4 \end{pmatrix}$$

Now the total cost of meals ordered on Saturday is

$$£(5 \times 3 + 10 \times 5 + 7 \times 4) = £93$$

and the total cost of meals ordered on Sunday is

$$£(12 \times 3 + 6 \times 5 + 9 \times 4) = £102$$

But $\mathbf{MP} = \begin{pmatrix} 5 & 10 & 7 \\ 12 & 6 & 9 \end{pmatrix} \begin{pmatrix} 3 \\ 5 \\ 4 \end{pmatrix} = \begin{pmatrix} 5 \times 3 + 10 \times 5 + 7 \times 4 \\ 12 \times 3 + 6 \times 5 + 9 \times 4 \end{pmatrix} = \begin{pmatrix} 93 \\ 102 \end{pmatrix}$

Hence the information giving the cost of meals ordered on Saturday and the cost of meals ordered on Sunday can be obtained from a matrix product,

i.e.
$$\mathbf{MP} = \begin{array}{c} \\ \text{Sat} \\ \text{Sun} \end{array} \begin{pmatrix} 93 \\ 102 \end{pmatrix} \begin{array}{c} £ \\ \\ \end{array}$$

Now consider the product **DM** where **D** $= (1 \quad 1)$,

i.e. $(1 \quad 1) \begin{pmatrix} 5 & 10 & 7 \\ 12 & 6 & 9 \end{pmatrix} = (5+12 \quad 10+6 \quad 7+9) = (17 \quad 16 \quad 16)$

Hence premultiplying **M** by $(1 \quad 1)$ effectively adds together the two entries in each column of **M**. Reference to **M** shows that this gives the total numbers of each meal ordered over the two days,

i.e.
$$\mathbf{DM} = (17 \quad 16 \quad 16) \begin{array}{c} \text{A} \quad \text{B} \quad \text{C} \\ \\ \end{array}$$

If we post multiply **M** by $\begin{pmatrix} 1 \\ 1 \\ 1 \end{pmatrix}$, this gives

$$\begin{pmatrix} 5 & 10 & 7 \\ 12 & 6 & 9 \end{pmatrix} \begin{pmatrix} 1 \\ 1 \\ 1 \end{pmatrix} = \begin{pmatrix} 22 \\ 27 \end{pmatrix}$$

This time it is the entries in each row of **M** that are added, so the result gives the total number of meals ordered on each of the two days,

i.e.
$$\begin{matrix} & & \text{No. of meals} \\ \text{Sat.} & \begin{pmatrix} 22 \\ & \\ 27 \end{pmatrix} \\ \text{Sun.} & \end{matrix}$$

EXERCISE 9b

1. A tennis club has three teams of players, team A, team B and team C. Each team plays in a league tournament and the results are displayed in the matrix **M** where

$$\begin{matrix} & & \text{Won} & \text{Drawn} & \text{Lost} \\ & \text{A} & 7 & 4 & 2 \\ \mathbf{M} = & \text{B} & 6 & 2 & 5 \\ & \text{C} & 5 & 3 & 5 \end{matrix}$$

If $\mathbf{N} = (1 \quad 1 \quad 1)$ and $\mathbf{P} = \begin{pmatrix} 1 \\ 1 \\ 1 \end{pmatrix}$

evaluate NM and MP and interpret the results.

Given that 2 points are awarded for a win, 1 point for a draw and no points for a lost match, evaluate $\mathbf{M} \begin{pmatrix} 2 \\ 1 \\ 0 \end{pmatrix}$ and interpret the result.

2. A furniture manufacturer makes three different types of table, A, B and C. The orders for each type of table for three separate months are displayed in the matrix **T** where

$$\begin{matrix} & & \text{I} & \text{II} & \text{III} \\ & \text{A} & 10 & 5 & 4 \\ \mathbf{T} = & \text{B} & 3 & 15 & 2 \\ & \text{C} & 4 & 10 & 10 \end{matrix}$$

The cost of raw materials for each type of table is given in the matrix **C** where

$$C = \begin{matrix} A \\ B \\ C \end{matrix} \begin{pmatrix} 10 \\ 15 \\ 12 \end{pmatrix} \quad £$$

and the time taken to make each type of table is given in the matrix **R** where

$$\text{Time (hrs)}$$
$$R = \begin{matrix} A \\ B \\ C \end{matrix} \begin{pmatrix} 5 \\ 4 \\ 5 \end{pmatrix}$$

Evaluate the following products and in each case interpret the result.

a) **TC**   b) **TR**

If the cost of labour in making any table is £10 per hour, evaluate **TC + 10TR** and interpret the result.

3. In one block of flats a milkman has three customers, A, B, and C. The matrix **M** shows the number of redtop, goldtop and silvertop bottles of milk ordered by each customer for one particular week. The matrix **C** shows the cost (in pence) of one bottle of each type of milk.

$$\text{If } M = \begin{matrix} A \\ B \\ C \end{matrix} \begin{pmatrix} 7 & 0 & 14 \\ 0 & 0 & 20 \\ 0 & 10 & 5 \end{pmatrix} \quad \text{and} \quad C = \begin{matrix} r \\ g \\ s \end{matrix} \begin{pmatrix} 26 \\ 27 \\ 25 \end{pmatrix} \quad \text{evaluate}$$

a) $(1 \ 1 \ 1)M$ b) $M \begin{pmatrix} 1 \\ 1 \\ 1 \end{pmatrix}$ c) **MC**   d) $(1 \ 1 \ 1)MC$

and in each case interpret the result.

4. A company has three factories, A, B and C. The matrix **E** shows the numbers of full-time, half-time and trainee employees at each factory. The matrix **W** shows the weekly wage paid to each category of employee.

$$\begin{matrix} & \text{Full-time} & \text{Part-time} & \text{Trainee} \\ E = \begin{matrix} A \\ B \\ C \end{matrix} & \begin{pmatrix} 10 & 4 & 3 \\ 8 & 2 & 6 \\ 15 & 12 & 5 \end{pmatrix} \end{matrix} \quad \text{and} \quad W = \begin{pmatrix} 120 \\ 50 \\ 60 \end{pmatrix} \begin{matrix} \text{Full-time} \\ \text{Part-time} \\ \text{Trainee} \end{matrix} \quad £$$

Evaluate   a) $E \begin{pmatrix} 1 \\ 1 \\ 1 \end{pmatrix}$ b) $(1 \ 1 \ 1)E$ c) **EW**   d) $(1 \ 1 \ 1)EW$

and in each case interpret the result.

5. The matrix **P** shows the numbers of men, women and children under 12 living in a hostel. The matrix **C** shows the daily calories required by a man, a woman and a child where

$$\begin{array}{ccc} \text{Men} & \text{Women} & \text{Children} \end{array}$$
$$\mathbf{P} = (\begin{array}{ccc} 10 & 30 & 45 \end{array}) \quad \text{and} \quad \mathbf{C} = \begin{array}{c} \text{Man} \\ \text{Woman} \\ \text{Child} \end{array} \begin{array}{c} \text{Calories} \\ \begin{pmatrix} 2000 \\ 1400 \\ 1000 \end{pmatrix} \end{array}$$

Find   a) **PC**   b) $\mathbf{P} \begin{pmatrix} 1 \\ 1 \\ 1 \end{pmatrix}$ and in each case interpret the result.

6. A garment manufacturer needs supplies of three items; machine yarn (y), zips (z), and buttons (b). These items are available from two sources, A and B. Matrix **Q** shows the quantities of each item needed for each of four yearly quarters. Matrix **C** shows the cost of each item from each of the two suppliers.

If

$$\mathbf{Q} = \begin{array}{c} y \\ z \\ b \end{array} \begin{pmatrix} \begin{array}{cccc} \text{I} & \text{II} & \text{III} & \text{IV} \\ 50 & 70 & 50 & 80 \\ 70 & 60 & 70 & 20 \\ 100 & 200 & 200 & 100 \end{array} \end{pmatrix} \qquad \mathbf{C} = \begin{array}{c} A \\ B \end{array} \begin{pmatrix} \begin{array}{ccc} y & z & b \\ 10 & 20 & 2 \\ 12 & 15 & 3 \end{array} \end{pmatrix}$$

evaluate   a) **CQ**   b) $\mathbf{CQ} \begin{pmatrix} 1 \\ 1 \\ 1 \\ 1 \end{pmatrix}$

and interpret the results. Which supplier is cheaper if the cost is considered
c) for the first quarter only   d) for the whole year?

7. A small firm employs three people, A, B and C. Matrix **W** shows the standard hourly rate (s.r.) and the overtime rate (o.r.) of each employee and matrix **T** shows the number of hours worked at the standard rate and the number of hours worked at the overtime rate for each employee for one week.

If $\mathbf{W} = \begin{array}{c} A \\ B \\ C \end{array} \begin{pmatrix} \begin{array}{cc} \text{s.r.}(\pounds) & \text{o.r.}(\pounds) \\ 3 & 4 \\ 2.5 & 3 \\ 4 & 6 \end{array} \end{pmatrix}$ and $\mathbf{T} = \begin{array}{c} \text{s.r.} \\ \text{o.r.} \end{array} \begin{pmatrix} \begin{array}{ccc} A & B & C \\ 36 & 30 & 10 \\ 5 & 4 & 1 \end{array} \end{pmatrix}$

evaluate **WT** and state what the figures x, y and z in the leading diagonal of **WT**, represent.

If Z is the matrix $\begin{pmatrix} x & 0 & 0 \\ 0 & y & 0 \\ 0 & 0 & z \end{pmatrix}$ evaluate $(\begin{array}{ccc} 1 & 1 & 1 \end{array})\mathbf{Z}\begin{pmatrix} 1 \\ 1 \\ 1 \end{pmatrix}$

and interpret the result.

8. The town Export has two mainline railway stations A and B. There are also four suburban railway stations C, D, E and F. Train services between A, B and C, D, E, F are provided, or not, as indicated by 1, or 0, in the matrix T, where

$$T = \begin{array}{c} \\ A \\ B \end{array} \begin{pmatrix} C & D & E & F \\ 1 & 0 & 1 & 0 \\ 1 & 1 & 0 & 1 \end{pmatrix}$$

Mainline services from stations A and B to the towns X, Y and Z, exist or not as indicated by 1 or 0 in the matrix S where

$$S = \begin{array}{c} X \\ Y \\ Z \end{array} \begin{pmatrix} A & B \\ 1 & 0 \\ 0 & 1 \\ 1 & 1 \end{pmatrix}$$

Find the product **ST** and interpret the result.

FINDING THE APPROPRIATE MATRIX

We have seen that premultiplying a matrix by an appropriate row of 1's, adds the entries in the columns of the matrix.

Similarly postmultiplying a matrix by an appropriate column of 1's, adds the entries in the rows of the matrix.

Consider again the matrices **M** and **P** listing the orders for different meals and the price of the meals, as introduced earlier in this chapter.

On page 175 we found the product MP where

$$MP = \begin{array}{c} \\ Sat \\ Sun \end{array} \begin{pmatrix} £ \\ 93 \\ 102 \end{pmatrix}$$

i.e. **MP** lists the cost of meals ordered on Saturday and the cost of meals ordered on Sunday.

The total cost of the meals over the weekend is obtained by adding the entries in **MP**. To find a matrix operation which does this, we need an operation which adds the entries in the column. This can be achieved by premultiplying **MP** by a row of 1's. As there are two entries in the column of **MP**, we need two entries in the row of 1's.
i.e. premultiplying by (1 1) will achieve the required result:

$$(1 \quad 1) \begin{pmatrix} 93 \\ 102 \end{pmatrix} = (195)$$

Now suppose that we want to find a matrix operation that will tell us the total number of meals ordered over the weekend.

Starting with $\mathbf{M} = \begin{pmatrix} 5 & 10 & 7 \\ 12 & 6 & 9 \end{pmatrix}$, we need an operation which will add the entries in the columns and then the entries in the rows (or vice-versa).

Premultiplying $\mathbf{M}$ by $(1 \ \ 1)$ adds the entries in the columns,

i.e. $(1 \ \ 1)\begin{pmatrix} 5 & 10 & 7 \\ 12 & 6 & 9 \end{pmatrix} = (17 \ \ 16 \ \ 16)$

then postmultiplying by a column of 1's (we need three 1's) will add the entries in $(17 \ \ 16 \ \ 16)$,

$$(17 \ \ 16 \ \ 16)\begin{pmatrix} 1 \\ 1 \\ 1 \end{pmatrix} = (49)$$

This example illustrates how matrices can be used to extract information from separate lists. We have chosen a very simple example to illustrate the methods, but its extension to an organisation handling many more items over a much longer time span is obvious, and in this case using a computer to handle the information has clear advantages.

EXERCISE 9c

1. Over the three terms of the school year a school needs incidental supplies of boxes of chalk, pads of file paper and pads of graph paper. If these items cost respectively £5, £1, and £1.50 each and if the columns of $\mathbf{R}$ represent the requirements of these items for each of the three terms, find a suitable row or column matrix which when multiplied by $\mathbf{R}$ will give the total cost of these items for each term.

$$\mathbf{R} = \begin{pmatrix} 1 & 2 & 1 \\ 0 & 5 & 2 \\ 2 & 10 & 5 \end{pmatrix} \begin{matrix} \text{Chalk} \\ \text{File paper} \\ \text{Graph paper} \end{matrix}$$

What matrix product needs to be evaluated to give the total cost for all these items for the year?

2. A shop stocks three brands of tea, X, Y and Z. The numbers of packets of each of these three brands that are sold in each of four consecutive weeks are shown in the matrix S, where

$$\mathbf{S} = \begin{matrix} \text{X} \\ \text{Y} \\ \text{Z} \end{matrix} \begin{matrix} \text{Wk 1} & \text{Wk 2} & \text{Wk 3} & \text{Wk 4} \\ \begin{pmatrix} 10 & 9 & 12 & 6 \\ 15 & 5 & 4 & 20 \\ 17 & 10 & 16 & 8 \end{pmatrix} \end{matrix}$$

Find the matrix operation which will give the number of each brand sold over the four week period.

The shopkeeper buys his supplies from either wholesaler A or wholesaler B. From A, packets of X, Y and Z cost 40 p, 30 p and 25 p each respectively. From B, packets of X, Y and Z cost 35 p, 30 p and 29 p each respectively. The shopkeeper wishes to compare the cost to him of replacing his stock of tea each week from the two wholesalers. Find the matrix product which will give him this information.

3. There are three secondary schools, A, B and C in Dovemouth. For administrative purposes each school is divided into upper, middle and lower schools. The matrix **F** gives the number of forms in each section of each school.

$$
\mathbf{F} = \begin{array}{c} \\ A \\ B \\ C \end{array} \begin{array}{ccc} \text{Upper} & \text{Middle} & \text{Lower} \\ \left(\begin{array}{ccc} 12 & 9 & 6 \\ 10 & 7 & 5 \\ 12 & 8 & 6 \end{array} \right) \end{array}
$$

In each school, there are 30 pupils in each lower school form, 28 pupils in each middle school form and 25 pupils in each upper school form.

Find the appropriate matrix products which will give the following information.

a) the number of forms in each school

b) the number of pupils in each school

c) the numbers of each of upper, middle and lower school forms in the town

d) the number of pupils in all three schools.

4. There is a nonstop train service between two towns, X and Y. There is one village, A, on the road between the two towns and country buses operate the following routes: a direct non-stop service between X and Y, a service from X, stopping at A, to Y and then directly back to X.

The numbers of public service connections between X, Y, and A are shown in the matrix **R** where

$$
\mathbf{R} = \begin{array}{c} \\ X \\ Y \\ A \end{array} \begin{array}{ccc} X & Y & A \\ \left(\begin{array}{ccc} 0 & 3 & 1 \\ 3 & 0 & 0 \\ 0 & 1 & 0 \end{array} \right) \end{array}
$$

a) Explain the significance of the rows and the columns of this matrix.

b) Find $\mathbf{R}^2$ (i.e. $\mathbf{R} \times \mathbf{R}$) and interpret the entries that appear in the main diagonal of $\mathbf{R}^2$.

5. The country Alphaland has two international airports, A_1 and A_2, and three provincial airports L, M and N. Internal flights between A_1, A_2 and L, M, N are shown by '1' if they exist, and by '0' if they do not exist, in the matrix **A**, where

$$\mathbf{A} = \begin{array}{c} \\ A_1 \\ A_2 \end{array} \begin{array}{c} \text{L} \quad \text{M} \quad \text{N} \\ \left(\begin{array}{ccc} 1 & 1 & 0 \\ 0 & 1 & 1 \end{array} \right) \end{array}$$

The country Betaland has two international airports, B_1 and B_2. The existence or otherwise or flights between Alphaland and Betaland is shown by the matrix **F** where

$$\mathbf{F} = \begin{array}{c} \\ B_1 \\ B_2 \end{array} \begin{array}{c} A_1 \quad A_2 \\ \left(\begin{array}{cc} 1 & 0 \\ 1 & 1 \end{array} \right) \end{array}$$

Betaland has three provincial airports X, Y and Z and the existence or otherwise of internal flights between B_2, B_2 and X, Y, Z is shown by the matrix **B** where

$$\mathbf{B} = \begin{array}{c} \\ X \\ Y \\ Z \end{array} \begin{array}{c} B_1 \quad B_1 \\ \left(\begin{array}{cc} 1 & 0 \\ 0 & 1 \\ 1 & 1 \end{array} \right) \end{array}$$

By finding the appropriate matrix product give the matrix which indicates the existence or otherwise of air routes between

a) the provincial airports in Alphaland and the international airports in Betaland

b) the provincial airports in Betaland and the international airports in Alphaland

c) the provincial airports of Alphaland and Betaland, and interpret the numbers other than 1 and 0 which appear.

10 GEOMETRIC PROOF

This chapter gives an idea of how to prove geometric properties and revises some of the properties covered in earlier books.

DEMONSTRATION BY DRAWING AND MEASUREMENT

Geometry is the study of the properties of figures. Most of the properties that we have looked at so far have been demonstrated and verified by drawing and measurement. For example, when we investigated the sum of the angles in a triangle we drew some triangles, measured the angles and added them up. The results, from triangles of several different shapes and sizes, always came to around 180°. From these results we concluded that the sum of the angles in *any* triangle is 180°.

This method could be called 'jumping to conclusions' and it is not satisfactory for many reasons. It is impossible to draw a line; if we could, we would not be able to see it because a line has no thickness. It is impossible to measure angles with absolute accuracy; the protractors used in schools are probably capable of measuring to the nearest degree. The only conclusion that can reasonably be drawn from our results is 'it *seems likely* that the angle sum of any triangle is 180°'. (The true result could really be 179.5°.) 'Proof' by demonstration of particular cases also leaves open the possibility that somewhere, as yet unfound, there lurks an exception to the rule.

DEDUCTIVE PROOF

Learning geometrical properties from demonstrations gives the impression that each property is isolated. However geometry can be given a logical structure where one property can be deduced from other properties. This forms the basis of deductive proof; we quote known and accepted facts and then make logical deductions from them.

For example, if we accept that
a) vertically opposite angles are equal
b) corresponding angles are equal,
then, using just these two facts, we can prove that alternate angles are equal.

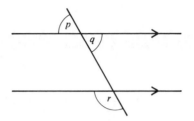

In the diagram $\hat{p} = \hat{q}$ (vertically opposite angles)

$\hat{p} = \hat{r}$ (corresponding angles)

$\Rightarrow \hat{q} = \hat{r}$

Therefore the alternate angles are equal.

The symbol $\Rightarrow$ means 'implies that' and indicates the logical deduction made from the two stated facts.

This proof does not involve angles of a particular size; p, q and r can be any size. Hence this proves that alternate angles are *always* equal whatever their size.

As a further example of deductive proof we will prove that, in *any* triangle, the sum of the interior angles *is* $180°$.

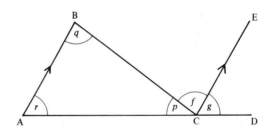

If $\triangle ABC$ is any triangle and if AC is extended to D and CE is parallel to AB then

$$\hat{p} + \hat{f} + \hat{g} = 180°$$ (angles on a st. line)   (1)

$$\hat{f} = \hat{q}$$ (alt. $\angle$'s)   (2)

$$\hat{g} = \hat{r}$$ (corr. $\angle$'s)   (3)

$$\Rightarrow \quad \hat{p} + \hat{q} + \hat{r} = 180°$$

i.e. the sum of the interior angles of *any* triangle is $180°$.

The statements above also lead to another useful fact about angles in triangles:

$$(2) \text{ and } (3) \quad \Rightarrow \quad \widehat{f} + \widehat{g} = \widehat{q} + \widehat{r}$$

i.e. an exterior angle of a triangle is equal to the sum of the two interior opposite angles.

Because this proof does not involve measuring angles in a particular triangle it applies to all possible triangles thus closing the loophole that there may exist a triangle whose angles do not add up to $180°$.

Notice how this proof uses the property proved in the first example, i.e. this proof follows the previous proof. The angle sum property of triangles can now be used to prove further properties.

Euclid was the first person to give a formal structure to Geometry. He started by making certain assumptions, such as 'there is only one straight line between two points'. Using only these assumptions (called axioms), he then proved some facts and used those facts to prove further facts and so on. Thus the proof of any one fact could be traced back to the axioms.

However when *you* are asked to give a geometric proof you do not have to worry about which property depends on which; you can use *any* facts that you know. One aspect of proof is that it is an argument used to convince other prople of the truth of any statement, so whatever facts you use must be clearly stated.

It is a good idea to marshal your ideas before starting to write out a proof. This is most easily done by marking right angles, equal angles and equal sides etc. on the diagram.

The exercises in this chapter give practice in writing out a proof.

For the next exercise the following facts are needed:

> vertically opposite angles are equal,
>
> corresponding angles are equal,
>
> alternate angles are equal,
>
> interior angles add up to $180°$,
>
> angle sum of a triangle is $180°$,
>
> an interior angle of a triangle is equal to the sum of the interior opposite angles,
>
> an isosceles triangle has two sides of the same length and the angles at the base of those sides are equal,
>
> an equilateral triangle has three sides of the same length and each interior angle is $60°$.

EXERCISE 10a

1.

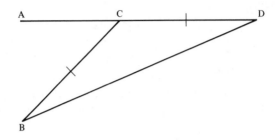

Prove that $\widehat{ACB} = 2\widehat{CDB}$.

2.

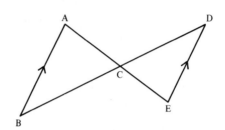

Prove that $\widehat{ACD} = \widehat{ABC} + \widehat{DEC}$.

3.

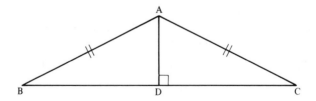

Prove that AD bisects BAC.

4.

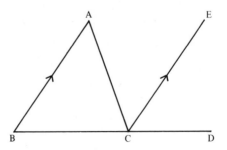

CE bisects $\widehat{ACD}$ and CE is parallel to BA. Prove that $\triangle ABC$ is isosceles.

5.

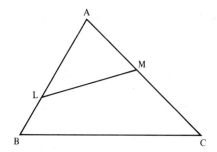

$A\widehat{M}L = A\widehat{B}C$. Prove that $A\widehat{L}M = A\widehat{C}B$.

6.

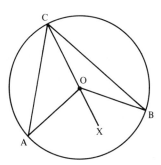

O is the centre of the circle.

a) Prove that $A\widehat{O}X = 2A\widehat{C}O$ b) Prove that $A\widehat{O}B = 2A\widehat{C}B$.

PARALLELOGRAMS, POLYGONS AND CONGRUENT TRIANGLES

The next exercise uses the following facts in addition to those already used.

In a parallelogram

> both pairs of opposite sides are parallel,
>
> both pairs of opposite sides are equal,
>
> both pairs of opposite angles are equal,
>
> the diagonals bisect each other.

To prove that a quadrilateral is a parallelogram we must show that it has *one* of the sets of properties listed.

Two triangles are congruent if

> either the three sides of one triangle are equal to the three sides of the other triangle,
>
> or two angles and one side of one triangle are equal to two angles and the corresponding side of the other triangle,
>
> or two sides and the included angle of one triangle are equal to two sides and the included angle of the other triangle,
>
> or if each triangle is right angled, the hypotenuse and one side of one triangle is equal to the hypotenuse and one side of the other triangle.

Only one of the above sets of properties need be established to prove the triangles congruent.

In a polygon

> the sum of the exterior angles is $360°$,
>
> the sum of the interior angles is $(180n - 360)°$ where n is the number of sides.

If the polygon is regular, all the sides are equal and all the interior angles are equal.

EXERCISE 10b

ABC is an isosceles triangle in which $AB = AC$. A point D is inside the triangle and $D\widehat{B}C = D\widehat{C}B$. Prove that AD bisects $B\widehat{A}C$.

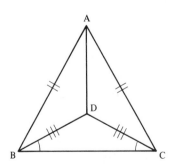

$$D\widehat{B}C = D\widehat{C}B \quad \text{(given)}$$

$\Rightarrow \qquad \triangle BCD$ is isosceles

$\Rightarrow \qquad BD = CD$

In $\triangle$s $\begin{array}{l}\text{ADB}\\\text{ADC}\end{array}$
$\qquad BD = CD \quad \text{(proved)}$
$\qquad AB = AC \quad \text{(given)}$
$\qquad AD$ is common

$\therefore \quad \triangle ADB$ and $\triangle ADC$ are congruent (SSS)

$\therefore \quad B\widehat{A}D = C\widehat{A}D$, i.e. AD bisects $B\widehat{A}C$.

1.

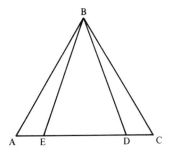

AEDC is a straight line. AB = BC and AE = DC.
Show that $\triangle AEB$ and $\triangle BDC$ are congruent. Hence prove that $\triangle BDE$ is isosceles.

2.

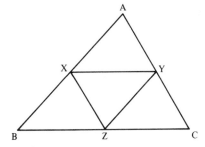

X, Y and Z are the midpoints of sides AB, AC and BC respectively.
Prove that $\triangle XYZ$ is congruent with $\triangle YZC$. Hence prove that BXYZ is a parallelogram.

3.

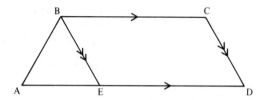

ABCD is a trapezium with BC parallel to AD and AB equal to CD. BE is parallel to CD and $C\widehat{D}E = 60°$. Prove that $\triangle ABE$ is equilateral.

4. ABCDEF is a regular hexagon. Prove that ABDE is a rectangle.

5.

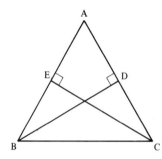

AB = AC,
BD is perpendicular to AC and CE is perpendicular to AB.
Prove that $\triangle BDC$ is congruent with $\triangle BEC$ and hence prove that $\triangle AED$ is isosceles.

6. ABCDEF is a hexagon in which AB is parallel and equal to ED and BC is parallel and equal to FE.
Join B to E and prove that $A\widehat{B}C = F\widehat{E}D$.
Hence prove that $\triangle ABC$ is congruent with $\triangle FED$.
Hence prove that ACDF is a parallelogram.

7.

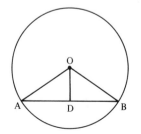

O is the centre of the circle and D is the midpoint of AB.
Prove that OD is perpendicular to AB.

8.

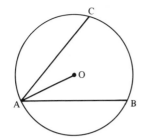

AB and AC are equal chords of a circle, centre O. Prove that AO bisects angle CAB.

CIRCLES

The next exercise introduces the use of the following facts.

The radius through the midpoint of a chord is perpendicular to the chord. (This was proved in question 7 of the last exercise.)

The angle subtended at the centre of a circle is equal to twice the angle subtended at the circumference by the same arc. (This is proved in question 6, Exercise 10a.)

All the angles subtended at the circumference by an arc of a circle are equal.

The angle in a semicircle is 90°.

The opposite angles of a cyclic quadrilateral add up to 180°.

The exterior angle of a cyclic quadrilateral is equal to the interior opposite angle.

EXERCISE 10c

1.

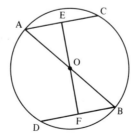

AB is a diameter and O is the centre of the circle. AC = BD and E, F are the midpoints of AC, BD. Prove that △AEO and △BFO are congruent. Hence prove that   a) EOF is a straight line   b) AC and DB are parallel.

2.

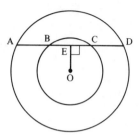

O is the centre of both circles (i.e. the circles are concentric). ABCD is a straight line and OE is perpendicular to ABCD. Prove that AB = CD.

3.

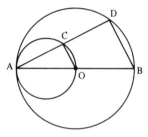

AB is a diameter and O is the centre of the larger circle. AO is a diameter of the smaller circle. ACD is a straight line. Prove that CO is parallel to DB.

4.

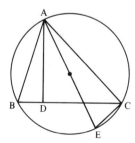

AE is a diameter of the circle and AD is perpendicular to BC. Prove that △AEC and △ABD are equiangular.

5.

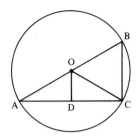

AOB is a diameter of the circle and OD bisects $A\widehat{O}C$. Prove that OD is parallel to BC.

6.

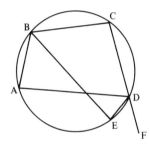

CDF is a straight line and BE bisect $A\widehat{B}C$. Prove that ED bisects $A\widehat{D}F$.

7.

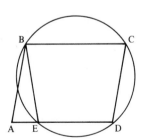

ABCD is a parallelogram. Prove that $B\widehat{A}E = B\widehat{E}A$.

11 CIRCLES AND TANGENTS

SECANTS AND TANGENTS

A straight line which cuts a circle in two distinct points is called a *secant*. The section of the line inside the circle is called a *chord*.

PQ is a secant and AB is a chord.

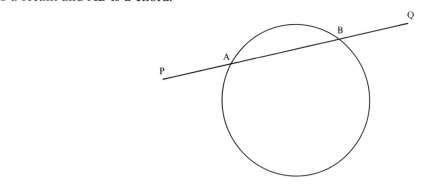

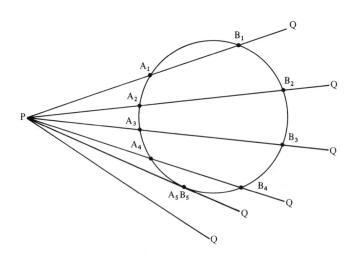

Imagine that the secant PQ is pivoted at P. As PQ rotates about P, we get successive positions of the points A and B, where the secant cuts the circle. As PQ moves towards the edge of the circle, the points A and B move closer together, until eventually they coincide.

194

When PQ is in this position it is called a *tangent* to the circle and we say that PQ touches the circle. (When PQ is rotated beyond this position it loses contact with the circle and is no longer either a secant or a tangent.)

We therefore define a tangent to a circle as a straight line which touches the circle.

The point at which the tangent touches the circle is called the point of contact.

PT is a tangent to the circle.
T is the point of contact.

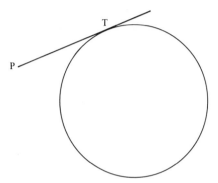

The *length of a tangent* from a point P outside the circle is the distance between that point and the point of contact. In the diagram the length of the tangent from P to the circle is the length PT.

EXERCISE 11a

1.

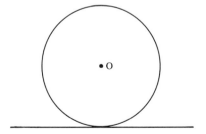

The diagram shows a disc, of radius 20 cm, rolling along horizontal ground. Describe the path along which O moves as the disc rolls. At any one instant,

a) how many points on the disc are in contact with the ground

b) how far is O from the ground

c) how would you describe the line joining O to the ground and what angle does it make with the ground?

2.

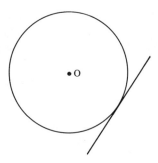

Copy the diagram and use a coloured or broken line to draw any line(s) of symmetry.

3.

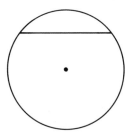

Copy the diagram and use a coloured or broken line to draw any line(s) of symmetry.

4.

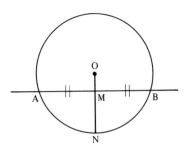

a) Show that the chord AB is perpendicular to the radius ON which bisects AB. (Join OA and OB.)

b) Now imagine that the chord AB slides down the radius ON. When the points A and B coincide with N, what has the line through A and B become? What angle does this line make with ON?

FIRST PROPERTY OF TANGENTS

The investigational work in the last exercise suggests that

> a tangent to a circle is perpendicular to the radius drawn from the point of contact.

The general proof of this property is an interesting exercise in logic. We start by assuming that the property is *not* true and end up by contradicting ourselves.

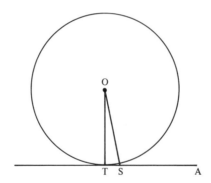

TA is a tangent to the circle and OT is the radius from the point of contact.

If we *assume* that $\widehat{OTS}$ is *not* 90° then it is possible to draw OS so that OS *is* perpendicular to the tangent, i.e. $\widehat{OST} = 90°$.

Therefore $\triangle OST$ has a right angle at S.
Hence   OT is the hypotenuse of $\triangle OST$
i.e.   OT > OS
∴   S is inside the circle, as OT is a radius.
∴   the line through T and S must cut the circle again.
But this is impossible, as the line through T and S is a tangent.
Hence the assumption that OTA ≠ 90° is wrong, i.e. $\widehat{OTA}$ *is* 90°.

EXERCISE 11b

Some of the questions in this exercise require the use of trigonometry.

The tangent from a point P to a circle of radius 4.2 cm is 7 cm long. Find the distance of P from the centre of the circle.

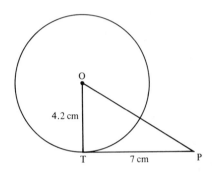

$$O\hat{T}P = 90°$$ (tangent perpendicular to radius)

$$OP^2 = OT^2 + TP^2$$ (Pythagoras' theorem)

$$= (4.2)^2 + 7^2$$

$$= 17.64 + 49$$

$$= 66.64$$

∴   OP = 8.16 correct to 3 s.f.

P is 8.16 cm from O.

In questions 1 to 8, O is the centre of the circle and AB is a tangent to the circle, touching it at A.

1.

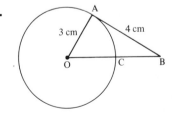

Find OB and CB.

2.

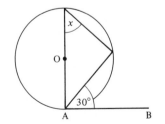

Find the angle marked x.

3.

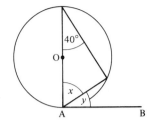

Find the angles marked x and y.

4.

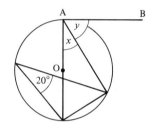

Find the angles marked x and y.

5.

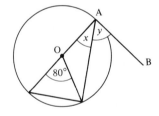

Find the angles marked x and y.

6.

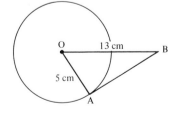

Find AB and $\widehat{OBA}$.

7.

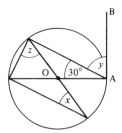

Find the size of the angles marked x, y and z.

8.

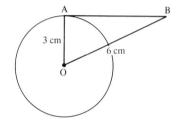

Find $\widehat{ABO}$.

9.

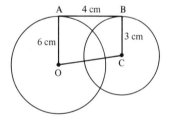

AB is a tangent to the circle with centre C, touching it at B. Find OC.

10.

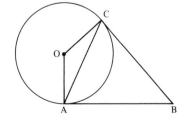

AB and BC are tangents to the circle touching it at C. Show that $\triangle ABC$ is isosceles.

11.

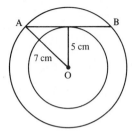

AB is a chord of the larger circle and a tangent to the smaller circle. If O is the centre of both circles, find the length of AB.

CONSTRUCTIONS

EXERCISE 11c

(Remember to draw a rough sketch before doing the construction.)

1. Draw a circle of radius 5 cm. Label the centre O and mark a point T on the circumference. Construct the tangent to the circle at T. (Use the fact that the radius OT is perpendicular to the tangent.)

2. Draw a circle of radius 4 cm. Label the centre of this circle C. Mark a point P distant 10 cm from C. Draw another circle on PC as diameter. Label the points where the two circles cut, A and B. What is the size of $\stackrel{\frown}{CAP}$? Describe the lines PA and PB in relation to the circle with centre C.

3. Draw a circle of radius 3 cm and mark a point P distant 6 cm from the centre of the circle. Use the method described in question 2 to construct the two tangents from P to the circle.

4.

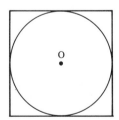

   a) The diagram shows a circle, centre O, inscribed in a square (i.e. the sides of the square are tangents to the circle). The radius of the circle is 2 cm. Find the length of a side of the square.

   b) Draw a square of side 8 cm. Construct the inscribed circle of the square.

PROOFS

EXERCISE 11d

AD is the diameter of a circle and AB is a tangent to the circle at A. BD meets the circle again at E and DE = EB.
Prove that $E\widehat{A}B = 45°$.

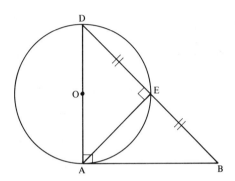

AD is a diameter

$\therefore$ $D\widehat{E}A = 90°$ (angle in semicircle)

In $\triangle AED$ and $\triangle AEB$

       DE = EB   (given)

AE is common

       $D\widehat{E}A = A\widehat{E}B$ (both 90°)

$\therefore$ $\triangle s \begin{matrix} AED \\ AEB \end{matrix}$ are congruent.   (SAS)

$\therefore$ $D\widehat{A}E = E\widehat{A}B$

But $D\widehat{A}B = 90°$ (angle between tangent
                                           and radius)

$\therefore$ $D\widehat{A}E = 45°$

---

1. AB is the diameter of a circle and D is a point on the circumference of the circle. A circle is drawn on AD as diameter. Prove that BD is a tangent to this circle.

2. A circle centre A is drawn to cut a circle, centre B, at points C and D such that $A\widehat{C}B = 90°$. Prove that AC is a tangent to the circle centre B.

3. AOB is a diameter of a circle, centre O. AD is a tangent to the circle at A and DB meets the circle again at C. Prove that $D\widehat{A}C = ABC$.

4. P is a point outside a circle with centre O. Tangents from P to the circle touch the circle at R and S. Prove that $\triangle ROP$ is congruent with $\triangle SOP$. Hence show that the tangents from P to the circle are equal in length.

5. AOB is a diameter of a circle centre O. AP is a tangent to the circle at A. A chord AC is drawn so that C and P are on the same side of AB. Prove that $C\widehat{A}P = A\widehat{B}C$.

SECOND TANGENT PROPERTY

The property proved about tangents in question 4 of the last exercise can be quoted, i.e.

> the two tangents drawn from an external point to a circle are the same length.

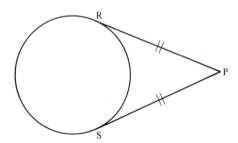

EXERCISE 11e

1.

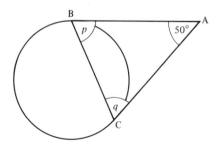

Find the sizes of the angles marked p and q.

2.

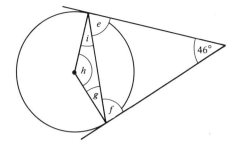

Find the sizes of the angles marked *e, f, g, h* and *i*.

3.

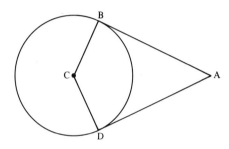

a) If $\widehat{BCD} = 130°$, find $\widehat{BAD}$.

b) What type of quadrilateral is ABCD ?

4.

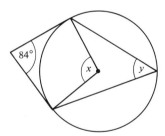

Find the sizes of the angles marked *x* and *y*.

5.

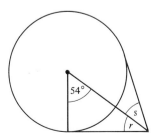

Find the sizes of the angles marked *r* and *s*.

6.

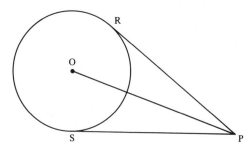

PR = 8 cm and OP = 10 cm. Calculate

a) the radius of the circle

b) the angle between the tangents.

7.

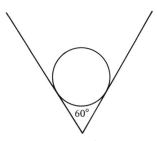

The diagram shows the cross-section through the centre of a ball placed in a hollow cone. The vertical angle of the cone is 60° and the diameter of the ball is 8 cm. Find the depth of the vertex of the cone below the centre of the ball.

8. A second ball is now placed in the cone described in question 7. If the diameter of the second ball is 20 cm, will it touch the first ball?

9.

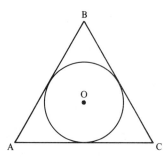

The circle, centre O, is inscribed in the equilateral triangle ABC. The sides of the triangle are each 20 cm long. Calculate the radius of the circle.

10. A circle of radius 4 cm is circumscribed by an equilateral triangle. Write down the angles between the sides of the triangle and the lengths of the lines joining the centre of the circle to the vertices of the triangle. Hence calculate the lengths of the sides of the triangle.

11. A circle of radius 4 cm is circumscribed by an isosceles right-angled triangle. Find the lengths of the sides of the triangle.

12. ABCD is a quadrilateral circumscribing a circle. If AC goes through the centre of the circle, prove that ABCD is a kite.

13. Construct a circle, centre O and radius 4 cm. Mark a point A on the circumference. Construct angle $\widehat{OAB} = 90°$ and hence draw the tangent AB. Mark any two points D and C on the circumference. Join A to D and A to C.
Measure $\widehat{CAB}$ and $\widehat{ADC}$. How do they compare ?

THIRD TANGENT FACT

Alternate segment theorem

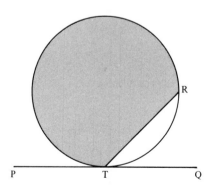

PQ is a tangent to the circle and TR is a chord. The major segment (which is shaded) is called the alternate segment with respect to the angle $\widehat{RTQ}$. Similarly the minor (unshaded) segment is alternate to the angle $\widehat{PTR}$.

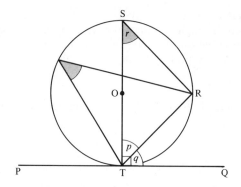

If TS is a diameter then

$$S\widehat{R}T = 90°$$ (angle in semi-circle)

$$S\widehat{T}Q = 90°$$ (angle between tangent and radius)

Now $\widehat{p} + \widehat{q} = 90°$

and $\widehat{p} + \widehat{r} = 90°$ (angles of $\triangle$)

$\Rightarrow$ $\widehat{q} = \widehat{r}$

But $\widehat{r}$ is equal to any angle subtended by the chord TR, i.e.

the angle between a tangent and a chord drawn from the point of contact is equal to any angle in the alternate segment.

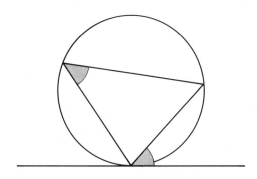

This result is known as the alternate segment theorem and can be quoted.

EXERCISE 11f

In questions 1 to 4, copy the diagram and shade the alternate segment with respect to the angle marked *x*.

1.

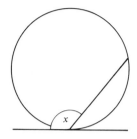

3.

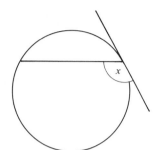

2.

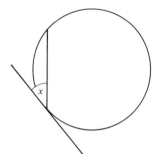

4.

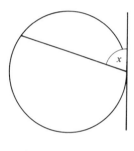

Find the size of the angles marked *x* and *y* in the diagram.

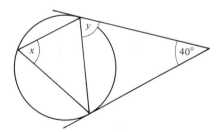

$y = 70°$ (base angle of isosceles triangle)

$x = y$ (alternate segment theorem)

$\therefore\ \ x = 70°$

Find the sizes of the angles marked by the letters.

5.

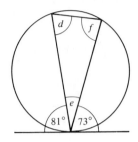

9.

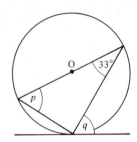

6.

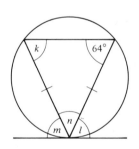

10.

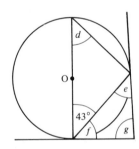

7.

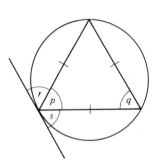

11.

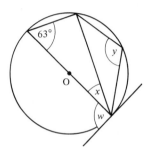

8.

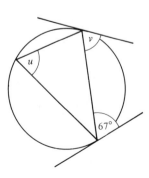

12.

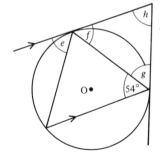

13.

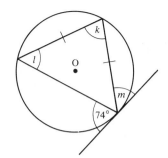

14.

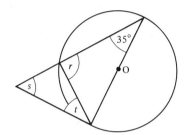

15.

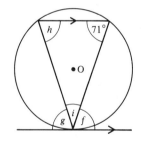

16.

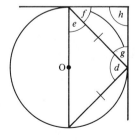

17.

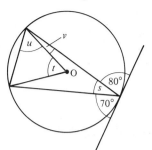

18.

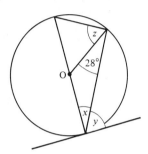

19.

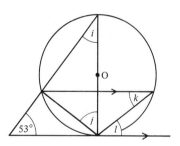

20.

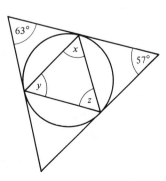

12 PROBABILITY

If we toss an unbiased die, each of the six numbers is equally likely to appear. We say that there are six equally likely *outcomes* to this experiment.

The probability of scoring 2 is $\frac{1}{6}$, since out of the six equally likely outcomes, only one is 'successful', i.e. is a 2.

We write $$P(2) = \tfrac{1}{6}$$

Similarly $$P(\text{even number}) = \tfrac{3}{6} = \tfrac{1}{2}$$

$$P(10) = 0$$

$$P(\text{either 1 or 2 or 3 or 4 or 5 or 6}) = \tfrac{6}{6} = 1$$

An event is often denoted by A.

If for example, A is the event 'throwing a 2'

then we write $$P(A) = \tfrac{1}{6}$$

The event 'throwing a number other than 2' is denoted by $\overline{A}$; hence $P(\overline{A}) = 1 - \tfrac{1}{6} = \tfrac{5}{6}$

Summary

$$P(\text{successful event}) = \frac{\text{number of successful outcomes}}{\text{total numbers of possible outcomes}}$$

$$P(\text{certainty}) = 1$$

$$P(\text{impossibility}) = 0$$

$$0 \leqslant P(A) \leqslant 1$$

$$P(\overline{A}) = 1 - P(A)$$

210

EXERCISE 12a

> If a card is drawn from a pack of 52 find the probability that it is
> a)   an ace
> b)   a card which is neither an ace nor a heart.
>
> a) $P(\text{ace}) = \frac{4}{52}$
>
> $$= \frac{1}{13}$$
>
> b) There are $(52 - 16)$ cards which are neither aces nor hearts.
>
> $P(\text{neither an ace nor a heart}) = \frac{36}{52}$
>
> $$= \frac{9}{13}$$

1.   A bag contains three red beads, two blue ones and four yellow ones. If one bead is drawn at random from the bag, find the probability that
a) the bead is red
b) the bead is red or blue
c) the bead is red or blue or yellow
d) the bead is green.

2.   A two-figure number is written down at random. Find the probability that
a) the number is greater than 72
b) the number contains at least one figure 6 (i.e. a number such as 62 or 16 or 66).

3.   A letter is picked at random from the English alphabet. Find the probability that
a) the letter is a vowel
b) the letter comes from the first half of the alphabet
c) the letter is one which appears in the word ALGEBRA.

4.   A two-figure number is written down at random. Find the probability that
a) the number is greater than 44
b) the number is less than 100.

A square board is marked as shown with a quadrant of a circle. A counter thrown on to it is equally likely to fall at any point on the board.

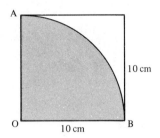

Find the probability that the counter falls in the shaded region

$$\text{Area of the square} = 10 \times 10 \, \text{cm}^2$$
$$= 100 \, \text{cm}^2$$
$$\text{Area of the quadrant} = \tfrac{1}{4}\pi r^2$$
$$= \tfrac{1}{4} \times \pi \times 100 \, \text{cm}^2$$
$$= 78.55 \, \text{cm}^2$$
$$P(\text{It falls in shaded region}) = \tfrac{78.55}{100}$$
$$= 0.786 \qquad \text{correct to 3 s.f.}$$

5.

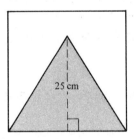

A square board of side 30 cm is marked as shown in the diagram with a triangle of height 25 cm.

When a coin is thrown on to the board its centre is equally likely to fall anywhere on the board. Find the probability that the centre of the coin falls in the shaded region.

The probability of drawing a red bead out of a bag of mixed beads is $\frac{2}{5}$. What is the probability of drawing a bead which is not red?

$$P(\text{non-red bead}) = 1 - \frac{2}{5}$$
$$= \frac{3}{5}$$

6. A drawer contains grey and brown socks. If a sock is picked at random the probability that it is grey is $\frac{3}{7}$, What is the probability of picking a brown sock?

7. A box contains yellow, red and black counters. A counter is drawn at random from the box. If the probability of getting a yellow counter is $\frac{1}{3}$ and of getting a red counter is $\frac{2}{5}$, find the probability of getting a black counter.

8. The probabilities of three hockey teams A, B and C winning a tournament are $\frac{1}{5}$, $\frac{1}{10}$ and $\frac{1}{2}$ respectively.
Find the probability that
a) either A or B will win
b) either A or B or C will win
c) none of these teams will win.

9. A coin is biased so that the probability of getting a tail is twice that of getting a head. What is the probability of getting a head?

In a studio audience if one person is picked at random the probability that it is a man is $\frac{5}{8}$. There are 240 people in the audience. How many are men?

$$\text{Number of men} = \frac{5}{8} \times 240$$
$$= 150$$

10. The probability of drawing an ace from a handful of 12 cards is $\frac{1}{6}$. How many aces are there in the hand?

11. In a car park there is a probability of $\frac{3}{8}$ that a car picked at random is British. There are 144 cars in the car park. How many of them are not British?

12. From a handful of red cards and black cards the probability of drawing a red card at random is $\frac{3}{5}$. There are 24 red cards. How many black cards are there?

13. On a shelf are French and German text books. The probability is $\frac{7}{13}$ that a book picked at random is French. There are 24 German books. How many French books are there?

14.

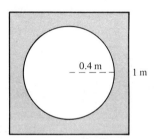

In a sideshow at a fête a disc is tossed on to a square board of side 1 m marked, as shown in the diagram, with a circle of radius 0.4 m. The centre of the disc is equally likely to fall at any point on the board.

Find the probability that the centre of the disc will fall on

a) the unshaded part   b) the shaded part.

Give your answers as decimals correct to three significant figures.

15. Forty pupils took an external examination. All but two of them took one or more of the subjects physics, chemistry and biology. The Venn diagram shows the number of candidates who entered for the various subject combinations.

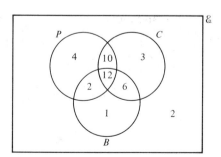

If one candidate is picked at random what is the probability that

a) the candidate took chemistry

b) the candidate took chemistry but not physics or biology?

c) If one of the physics candidates is picked at random, what is the probability that this candidate took both biology and chemistry as well?

16. A two-digit number is to be formed by choosing two different digits from 1, 2, 3 and 5 without repetition. List the numbers that can be formed in this way. If the digits are picked at random in this way what is the probability of forming a number which is divisible by 5?

In questions 17 to 20, several alternative answers are given. Write down the letter that corresponds to the correct answer.

17.

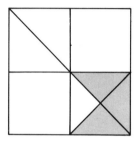

A counter is thrown onto a board which is marked as shown in the diagram. Its centre is equally likely to fall at any point on the board.
The probability that it lands on the shaded area is

A $\frac{1}{5}$ **B** $\frac{3}{10}$ **C** $\frac{13}{16}$ **D** $\frac{3}{16}$

18. A bag contains white counters and black counters. There are 24 white counters. The probability of drawing a black counter at random is $\frac{5}{8}$.
The number of black counters in the bag is

A 15   **B** 40   **C** 64   **D** 35

19. A card is drawn at random from a pack of 52 cards.
The probability that the card is neither a king nor black is

A $\frac{6}{13}$ **B** $\frac{1}{2}$ **C** $\frac{12}{13}$ **D** $\frac{7}{13}$

20. A disc is drawn from a bag containing 30 discs numbered to 1 to 30.
A small prize is given if a prime number is drawn.
The probability of winning a prize is

A $\frac{11}{30}$ **B** $\frac{1}{3}$ **C** $\frac{2}{5}$ **D** $\frac{3}{10}$

TWO EVENTS ━━

When two coins are tossed, the set of possible outcomes is

$$\{HH, HT, TH, TT\}$$

Each outcome is equally likely so we can see that $P(2\text{ heads}) = \frac{1}{4}$.

By listing the possible equally likely outcomes the required probability can be found. However, if two dice are tossed the list of possibilities is long and confusing. We need an organised way of setting out the list so that there is no risk of missing any outcome. This is considered in the next section.

POSSIBILITY SPACES ━━

We list the outcomes in the following table using crosses.

First die

		1	2	3	4	5	6
Second die	1	×	×	×	⊗	×	×
	2	×	×	⊗	×	×	☒
	3	×	⊗	×	×	☒	×
	4	⊗	×	×	☒	×	×
	5	×	×	☒	×	×	×
	6	×	☒	×	×	×	×

The table shows that there are 36 possible outcomes.

To find the probability of getting a total score of 5 we ring the crosses that mark the outcomes such as $1+4$, $2+3$ and so on.

Hence
$$P(5) = \tfrac{4}{36} = \tfrac{1}{9}$$

Similarly, we can find the probability of scoring, say, 8. We put squares around the outcomes that total 8.

Hence
$$P(8) = \tfrac{5}{36}$$

EXERCISE 12b

In each question from 1 to 4, draw a possibility space to show the outcomes when two dice are tossed and use it to find the required probabilities. Use different marks such as a ring and a square, or different colours, for the first two parts of each question.

1. Find the probability that
a) the sum of the two numbers is 6
b) the difference between the two numbers is 2
c) the sum is 6 and the difference is 2.

2. Find the probability that
a) prime numbers appear on both dice
b) at least one prime number appears
c) only one prime number appears.

3. Find the probability that
a) the sum of the two numbers is 8 or more
b) the difference between the two numbers is 2 or less
c) the sum of the two numbers is 8 or more and their difference is 2 or less.

4. Find the probability that
a) an even number appears on both dice
b) an odd number appears on both dice
c) an even number greater than 2 appears on both dice.

5. A four-sided spinner has the numbers 1 to 4 marked on it. It is spun twice and the two scores are noted. Draw a possibility space table to show the outcomes.
Find the probability that
a) the total score is even
b) the two separate scores are both even
c) the product of the scores is even.

6. I have two bags each containing four hyacinth bulbs and I know that each contains a pink, a blue, a yellow and a white bulb. If I take one bulb at random from each bag, find the probability that
a) the hyacinths will be the same colour
b) the hyacinths will be of different colours.

INTRODUCTION TO A NEW TYPE OF PROBLEM

In the previous exercise the second event (e.g. tossing a second die) is not affected by what happened first. In the following exercise, however, the probability of the second event varies depending on what happened first.

Possibility spaces cannot be used in this exercise.

EXERCISE 12c

A card is drawn from a pack of 52 playing cards.
a) What is the probability of drawing a red card?
b) If the first card is red and is not replaced what is the probability that a second card drawn is red?

a)   There are 26 red cards out of 52,
$$\text{therefore} \quad P(\text{red card}) = \tfrac{26}{52} = \tfrac{1}{2}$$

b)   There are now 25 red cards left out of 51 cards,
$$\text{therefore} \quad P(\text{2nd red card}) = \tfrac{25}{51}$$

1. A bag contains 5 red beads and 3 blue beads.
a) What is the probability of drawing   i) a red bead   ii) a blue bead?
b) If a red bead is drawn first and is not replaced, what is now the probability of drawing a blue bead?
c) If a red bead is drawn first and not replaced, what is the probability of drawing a second red bead?

2. A card is drawn from a pack of 52 cards.
a) Give the probability of drawing
   i) a nine   ii) a heart
b) If a heart is drawn and not replaced, what is now the probability of drawing a heart?
c) If a nine is drawn first and not replaced, what is now the probability of drawing a ten?

3. A hutch contains 4 white and 5 grey guinea pigs. When the door is opened they come out in random order.

a) Give the probability that the first out is white.

b) If the first out is white, what is the probability that the second out is
   i) white   ii) grey?

c) If the first out is grey, what is the probability that the second out is
   i) white   ii) grey?

PROBABILITY TREES

Suppose that we have seven cards of which two are red and five are black. When a card is drawn at random the probability that it is a red card is $\frac{2}{7}$ and that it is a black card is $\frac{5}{7}$. We can show this in a diagram.

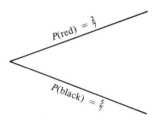

If we have drawn a red card, there are five black cards and one red card left. The probabilities of drawing a red or a black card are now $\frac{1}{6}$ and $\frac{5}{6}$.

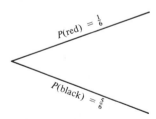

Similarly if a black card is drawn first, the diagram showing the probabilities for the colour of the second card is

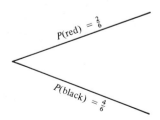

These three diagrams can be combined in one diagram.

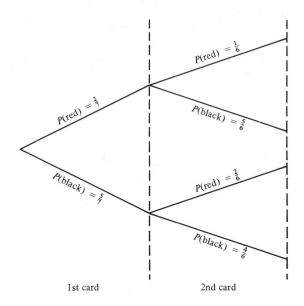

1st card   2nd card

If two cards are drawn out on a large number of occasions, say 420 times (starting afresh with the seven cards each time), we would expect to get a red card first on $\frac{2}{7} \times 420$, i.e. 120 occasions. On $\frac{1}{6}$ of these 120 occasions, i.e. on 20 occasions we would expect to get a second red card

$$\therefore \qquad\qquad P(\text{2 red cards}) = \tfrac{20}{420} = \tfrac{1}{21}$$

Note that $\frac{1}{21}$ is also given by $\frac{2}{7} \times \frac{1}{6}$

This result can be obtained from the tree diagram. Follow a path along the required branches (i.e. red card first, red card second), giving an outcome of two red cards, and *multiply* together the probabilities on the two branches.

Similarly $\qquad\qquad P(\text{2 black cards}) \qquad = \tfrac{5}{7} \times \tfrac{4}{6} = \tfrac{10}{21}$

and $\qquad\qquad P(\text{red first, black second}) = \tfrac{2}{7} \times \tfrac{5}{6} = \tfrac{5}{21}$

We assume that, unless stated otherwise, the first card is not replaced before the second card is drawn.

EXERCISE 12d

In this exercise, the first object drawn is not replaced before the second object is drawn.

A box contains ten counters; six are black and four are white. Two counters are drawn at random. Find the probability that
a) both are black   b) both are white.

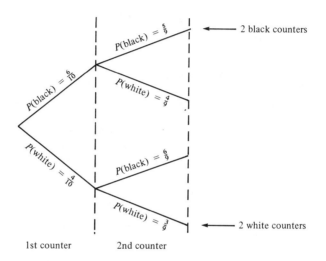

- 2 black counters
- 2 white counters

1st counter   2nd counter

a) $P(2\,\text{black}) = \frac{6}{10} \times \frac{5}{9} = \frac{1}{3}$

b) $P(2\,\text{white}) = \frac{4}{10} \times \frac{3}{9} = \frac{2}{15}$

1. A bag contains four green beads and five yellow beads. Two beads are withdrawn at random.

a) Find the probability that the first bead is green.

b) If the first bead is yellow find the probability that the second bead is green.

Draw a probability tree to show the probabilities when two beads are withdrawn and find the probability that

c) both beads are green

d) the first bead is yellow and the second is green.

2. A hand of ten cards contains four hearts and six clubs.
Two cards are drawn at random from the hand.

a) What is the probability that the first card is a heart?

b) If the first card is a heart what is the probability that the second card is a heart?

c) Draw a probability tree and find the probability that both cards are clubs.

3. Seven cards are numbered 1 to 7 and two cards are drawn at random. Draw a probability tree to show the probabilities of drawing odd or even cards. Find the probability that

a) the first card is even

b) both cards are even

c) both cards are odd

d) the first card is even and the second is odd

e) the first card is odd and the second is even

f) one card is odd and one even in any order.
(Use the answers to (d) and (e) to answer (f).)

4. a) If a drawing pin is dropped, it is three times as likely to land point up as point down. What is the probability that it will land point up?

b) Two such drawing pins are dropped. What is the probability that both will land point up?

5. A birdcage contains six blue and three green budgerigars. When the door is opened the birds come out one at a time in random order.

a) What is the probability that the first bird is blue?

b) If the first bird is blue, what is the probability that the second bird is blue?

Find the probability that

c) one of the first two birds is blue and one green

d) the first three birds out are all blue.

SIMPLIFIED TREE DIAGRAMS

We do not always have to draw every possible branch of a tree diagram. If, for instance, in throwing a die three times, we are interested only in the number of 6s thrown, the tree diagram need have only two branches per throw. One branch is for 'throwing a 6', the other is for 'throwing a number other than 6'.

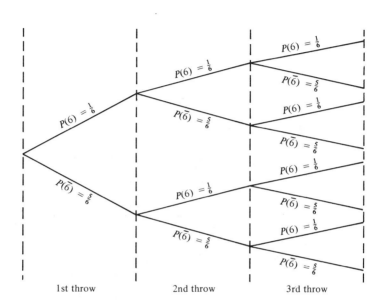

| 1st throw | 2nd throw | 3rd throw |

ADDITION OF PROBABILITIES

We have seen that the required outcomes are sometimes given by following more than one path along the branches.

We *add* the probabilities resulting from each path; each new possible way of achieving the event increases the probability.

MULTIPLICATION AND ADDITION

We *multiply* the probabilities when we follow a path along the branches of the probability tree and *add* the results of following several paths.

EXERCISE 12e

In this exercise, the first object drawn is not replaced before the second object is drawn.

Two dice are tossed. What is the probability of getting a prime number on one die and 1 on the other?

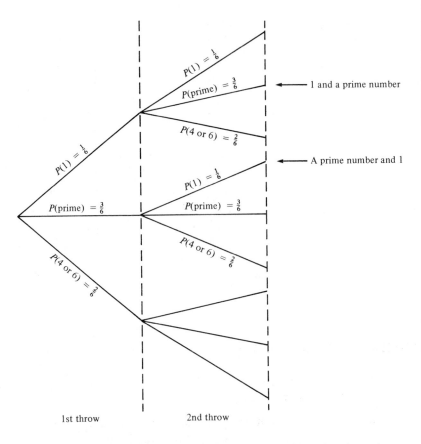

$$P(\text{a prime number and 1 in any order}) = \left(\tfrac{1}{6} \times \tfrac{3}{6}\right) + \left(\tfrac{3}{6} \times \tfrac{1}{6}\right)$$

$$= \tfrac{1}{12} + \tfrac{1}{12}$$

$$= \tfrac{2}{12}$$

$$= \tfrac{1}{6}$$

1. A hand of cards contains eight cards of which five are hearts and three are spades. Two cards are drawn at random. Draw a tree diagram to show the probabilities and find the probability that

    a) one card is a heart and one is a spade

    b) the two cards belong to the same suit.

2. Simon has six grey socks and four white ones in a drawer. He takes out two socks in the dark.
    What is the probability that they are of different colours?

3. The probability that the weather is fine on Monday is $\frac{1}{3}$. If it is fine, the probability that I can get on my bus is $\frac{3}{4}$. If it is not fine the probability that I can get on my bus is $\frac{1}{4}$.
    Find the probability that I can get on my bus on Monday.

4. Three coins are tossed. Find the probability of getting

    a) three heads   b) a head and two tails.

5. Two bags of hyacinth bulbs each contain four bulbs, a pink, a blue, a yellow and a white. If I take one bulb at random from each bag, find the probability that both bulbs are

    a) pink   b) blue   c) yellow   d) white   e) the same colour.

    Compare this method of answering (e) with the method used in Exercise 12b, question 6(a).

6. A box contains eight hard-centred and nine soft-centred chocolates. Two are selected at random. What is the probability that one is hard-centred and one is soft-centred?

7. The probability that Mr Brown completes a crossword puzzle is $\frac{2}{3}$, that Mrs Black completes it is $\frac{1}{2}$ and that Mr White completes it is $\frac{1}{3}$.
    Find the probability that

    a) all three complete it

    b) just two out of the three solve it.

8. The probability that Mr. Brodie, on his way to work, has to stop at the first set of traffic lights is $\frac{2}{5}$, and that he has to stop at the second set of traffic lights is $\frac{1}{3}$. Find the probability that he has to stop at just one of the sets of traffic lights.

HARDER PROBLEMS

EXERCISE 12f

1. In a test a group of one hundred pupils were given marks out of 60.
The table shows the number achieving the various marks

Marks	1–10	11–20	21–30	31–40	41–50	51–60
Frequency	4	16	20	24	27	9

a) State the probability that a pupil chosen at random will have a mark
   i)   from 41 to 50 inclusive
   ii) greater than 50
   iii) 20 or less.

b) A second group of one hundred pupils were tested and ten scored more
   than 50 marks. If one pupil is chosen at random from each of the groups,
   find the probability that
   i) both will have scored more than 50
   ii) just one will have scored more than 50.

2. Each of the following draws is from a set of four cards which are numbered
1, 3, 6, 8.

a) One card is drawn at random. Find the probability that the number on
   the card is a prime number.

b) Two cards are drawn at random. Find the probability that the numbers
   on both cards are multiples of 3.

c) Two cards are drawn at random. Find the probability that the sum of the
   two numbers is 9.

3. At a fête, thirty tickets numbered 1 to 30 are placed in a drum. A ticket
drawn from the drum wins a prize if its number is a multiple of 5. A ticket,
once drawn, is not replaced. Two tickets are drawn in succession. Find the
probability that

a) the first ticket wins a prize

b) the first ticket does not win a prize but the second does.

Three tickets are drawn in succession.

c) What is the probability that no ticket wins a prize?

d) What is the probability that at least one ticket wins a prize?

4. A biased die is such that the probability of throwing a six is $\frac{1}{3}$, and the
probability of scoring each of the other numbers is $\frac{2}{15}$.
I have two biased dice and two fair dice.
If I throw   a) two biased dice   b) one fair die and one biased die,
calculate the probability of obtaining a total score of   i) 12   ii) 11.

5. In a sideshow at a fête, a player is required to roll balls towards five channels marked with the scores 1 to 5.

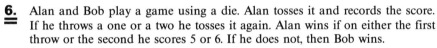

The probabilities of achieving the various scores are

$$P(1) = \tfrac{1}{10}$$
$$P(3) = \tfrac{1}{5}$$
$$P(5) = \tfrac{2}{5}$$
$$P(4) = \tfrac{1}{5}$$
$$P(2) = \tfrac{1}{10}$$

With two balls, find the probability of achieving a total score of

a) 10   b) 3   c) 4

With three balls, find the probability of achieving a total score of

d) 15   e) 3

6. Alan and Bob play a game using a die. Alan tosses it and records the score. If he throws a one or a two he tosses it again. Alan wins if on either the first throw or the second he scores 5 or 6. If he does not, then Bob wins.

a) What is the probability that Alan wins?

b) Who is more likely to win, Alan or Bob?

13 LOCI AND CONSTRUCTIONS

LOCI IN TWO DIMENSIONS

A locus is the set of all the points whose positions satisfy a given rule.

When the locus is a straight or curved line it is convenient to think of the locus as the path that is traced out by a single moving point.

Remember that every point on a locus must obey the given conditions or law, and that every point which obeys the law must lie on the locus.

The plural of locus is loci.

EXERCISE 13a

Describe, and illustrate with a sketch, the locus of the tip of the minute hand of a clock as it moves between 12 noon and 12.30 p.m.

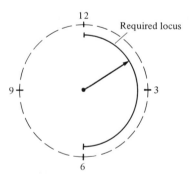

The required locus is a semicircle, centre at the centre of the clockface, radius the length of the minute hand.

In questions 1 to 10 describe, and illustrate with a sketch, the given locus.

1. The tip of the minute hand of a clock as it moves between 1 a.m. and 2 a.m.

2. The tip of the hour hand of a clock as it moves between 1 a.m. and 2 a.m.

3. A cricket ball when bowled at the wicket.

4. A cricket ball when hit along the ground for four.

5. A cricket ball when hit for six.

6. The centre of the wheel of a bicycle as the bicycle
  a) travels in a straight line
  b) travels around a bend.

7. The number at the top of this page as you turn the page over.

8. A satellite circling the earth.

9. The earth moving around the sun.

10. A goat on the end of a rope winding it around a tree.

A rod OA turns through a complete revolution about one end, O, which is fixed. Describe the locus of A.

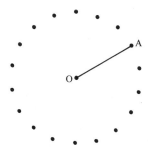

(Mark several possible positions for A until the overall shape of the path of A becomes clear.)

The locus of A is the circumference of a circle, centre O radius OA.

11. The minute hand of a clock is 80 cm long. Describe the locus of its tip
a) from 5 a.m. to 6 a.m.   b) from 2.15 p.m. to 2.45 p.m.

12. Describe the locus of a point on this page which moves so that it is always 3 cm from the top edge of this page.

13. Draw a straight line AB on a page of your exercise book. Describe the locus of a point X on the page which moves so that it is always 3 cm away from AB.

14.

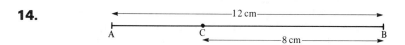

The rod AB is rotated about C. Describe the locus of
a) the point A   b) the point B.

15.

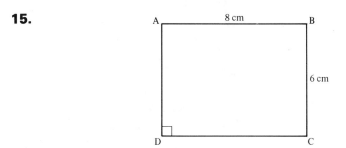

ABCD is a rectangle in which AB = 8 cm and BC = 6 cm. Describe the locus of points that are 3 cm from both AB and DC.

16.

A and B are two fixed points. Describe the locus of the points on this page that are equidistant from A and B.

17.

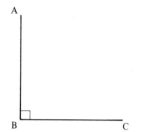

If $\widehat{ABC} = 90°$ describe the locus of points that are equidistant from AB and BC.

18. ABCD is a square of side 10 cm. Find the locus of points within the square equidistant from
a) AB and BC   b) AB and AD.
Is there any point that is equidistant from all three lines AB, BC and AD?
If so, where is it?

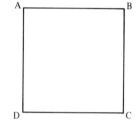

19.

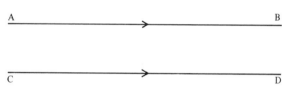

AB and CD are two parallel lines. Describe the locus of a point, between the two lines, that is always twice as far from AB as it is from CD.

20.

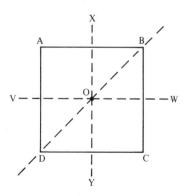

ABCD is a square, centre O, of side 4 cm. VW is parallel to AB and XY is parallel to AD. Describe the locus of A as the square is rotated about
a) XY   b) VW   c) DB   d) the axis through O perpendicular to ABCD.

SPECIAL LOCI

In questions 11 and 12 from exercise 13a the locus is a set of points traced out by a particular point, while in questions 13–19 the locus is a set of position points all of which exist at the same time.

These loci have introduced us to the four most important loci in two dimensional work.

1. The locus of a point that moves in such a way that it is always at a fixed distance from a fixed point is called a circle. The fixed point is the centre of the circle, and the fixed distance is its radius.

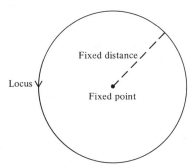

2. The locus of a point that moves in such a way that it is at a constant distance (d) from a line through two fixed points A and B, is the pair of straight lines drawn parallel to AB and distant d from it.

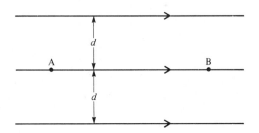

3. For points that are equidistant from two fixed points A and B, the locus is the perpendicular bisector of AB.

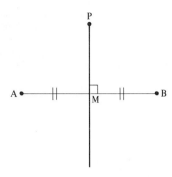

4. For points that are equidistant from two intersecting straight lines AXB and CXD, the locus is the pair of bisectors of the angles between the given lines. These bisectors are always at right angles to each other.

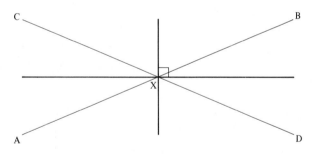

EXERCISE 13b

1.

M is the midpoint of a chord AB of fixed length in a circle, centre O. Describe the locus of M as AB moves around the circle.

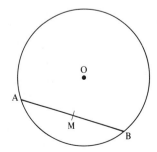

2.

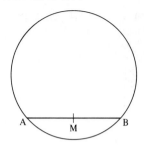

M is the midpoint of a chord AB. Describe the locus of the midpoints of the set of chords parallel to AB.

3.

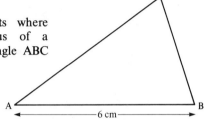

A and B are two fixed points where AB = 6 cm. Describe the locus of a third point C if the area of triangle ABC is 12 cm².

4.

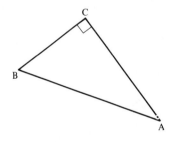

A and B are two fixed points. Describe the locus of C if $A\widehat{C}B = 90°$.

5. A is a fixed point. Describe the locus of the centres of circles which pass through A and have a radius of 5 cm.

6.

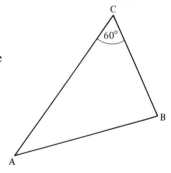

A and B are two fixed points. Describe the locus of C if $A\widehat{C}B = 60°$.

(Remember that angles in the same segment of a circle are equal.)

7.

A is a point on OY such that OA = 4 cm and B is a point on OX. If X$\widehat{O}$Y = 90° describe the locus of the midpoint of AB as B moves along OX.

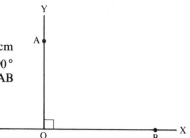

8.

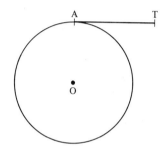

TA is a tangent, of fixed length, to a circle centre O. If A is the point of contact, describe the locus of T as A moves around the given circle.

9.

Sketch the locus of   a) D   b) C as the rectangle ABCD is rotated through 90° clockwise about A in the plane of the page.

10.

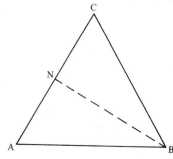

ABC is an equilateral triangle. The triangle is rotated clockwise about B until BC becomes parallel to the lower edge of the page. Sketch the locus of   a) C b) the foot, N, of the perpendicular from B to AC.
What angle has BA turned through ?

11.

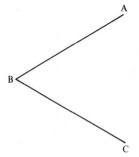

Draw the locus of the centres of circles that touch both AB and BC.

12.

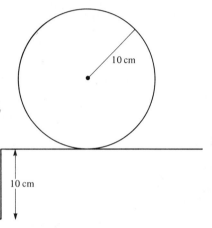

A wheel of radius 10 cm rolls across a horizontal path, and then down a step 10 cm deep, before continuing to roll horizontally. Sketch the locus of the centre of the wheel.

13.

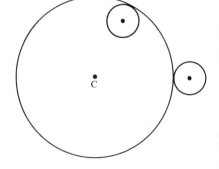

Describe the locus of the centre of a coin of diameter 2 cm if it
a) rolls around the inside of a circle
   with centre C and radius 5 cm
b) rolls around the outside of a circle
   of radius 5 cm.

14. Describe the locus of the centre of a circle, of variable radius, which passes through two fixed points A and B.

15. Draw any triangle ABC such that $\widehat{ABC} = 90°$. Draw the locus of points equidistant from   a) A and B   b) B and C.

What do you notice about the point of intersection of the two loci?

16.

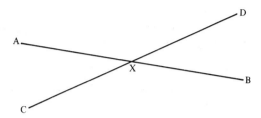

A house P is to be built 100 m from a road AXB and 50 m from a road CXD. Show on a sketch the possible positions for P.

17.

X and Y represent two houses 100 m apart. Sketch the loci that will enable you to shade the area of land that is both nearer to X than to Y and is within 60 m of Y.

18.

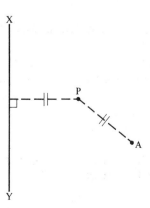

Plot the locus of P if P is a point whose distance from a fixed point A is equal to its distance from a fixed line XY.

CONSTRUCTIONS

EXERCISE 13c

Remember to make a rough sketch before you start each construction.

1. Construct an angle $\widehat{ABC}$ of $60°$. Construct the locus of points equidistant from AB and BC.

2. A and B are two points such that $AB = 9\,cm$. Construct the locus of points equidistant from A and B.

3. A and B are two points on a straight line of indefinite length, such that $AB = 8\,cm$.
a) Construct the locus of points that are $3\,cm$ from A.
b) Construct the locus of points that are $2\,cm$ from the line AB.
c) In how many points do these loci intersect? How far is each point from B?

4. Two straight lines AXB and CXD intersect at X such that $A\widehat{X}C = 90°$.
a) Construct the locus of points that are $4.5\,cm$ from X.
b) Construct the locus of points equidistant from AXB and CXD.
c) In how many points do these loci intersect? How far is each point from X?

5. Construct a triangle ABC in which $AB = 10\,cm$, $AC = 9\,cm$ and $BC = 8\,cm$.
a) Construct the locus of points equidistant from A and B.
b) Construct the locus of points equidistant from AB and AC.
c) Describe the point of intersection of the loci you have drawn in (a) and (b).

6. Draw a line AB that is $10\,cm$ long. Construct the locus of a point P such that $\triangle ABP$ is isosceles.

7. Draw a line AB that is $8\,cm$ long. Construct the locus of a point C such that $A\widehat{C}B = 90°$ (C may lie on either side of AB).

8. a) Construct a rectangle ABCD such that AB = 12 cm and BC = 8 cm.

  b) Draw the locus of points equidistant from AB and BC.

  c) Draw the locus of points equidistant from A and B.

  d) Mark the point P that lies on the loci referred to in both (b) and (c). Measure PC.

9. a) Draw AB of length 12 cm. Construct the locus of a point P, above AB, such that $\widehat{APB} = 90°$.

  b) Draw the locus of points that are 5 cm above AB.

  c) Mark the points P and Q that are 5 cm from AB, such that $\widehat{APB} = \widehat{AQB} = 90°$. Find the difference in length between AP and PB.

10. a) Construct a rectangle ABCD such that AB = 10 cm and BC = 8 cm.

  b) Draw the locus of points equidistant from AB and CD.

  c) Draw the locus of points, within the rectangle, that are 8 cm from C.

  d) Mark the point E, that is both equidistant from AB and CD, and 8 cm from C. Measure AE.

11. Draw a line AB that is 8 cm long. Construct the locus of a point P such that the area of △ABP is 24 cm².

12. a) Construct a triangle ABC in which AB = 12 cm, AC = 11 cm and BC = 8 cm.

  b) Draw the locus of points equidistant from AB and AC.

  c) Draw the locus of points equidistant from AC and CB.

  d) Mark, with I, the point of intersection of the two loci you have found in (b) and (c). What is special about the point I?

13. a) Construct a triangle ABC in which AB = 13 cm, $\widehat{ABC} = 45°$ and $\widehat{BAC} = 30°$.

  b) Draw the locus of points that are 2.5 cm from BC.

  c) Draw the locus of points that are 1.5 cm from AB.

  d) Hence find the point, P, within the triangle, that is 2.5 cm from BC and 1.5 cm from AB. Measure AP.

Construct a triangle ABC in which AB = 9.5 cm, BC = 7 cm and ABC = 60°. Find the point D, within the triangle, that is 2 cm from AB and 5 cm from C. Measure BD.

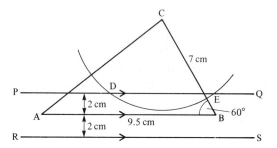

(Points that are 2 cm from AB lie on one or other of the two lines, PQ and RS, that are shown parallel to AB. Points that are 5 cm from C lie on the circle, centre C, radius 5 cm. This circle cuts PQ at D and E but cannot cut RS. Therefore RS need not be drawn in the accurate construction. From the sketch, D satisfies the given conditions but we cannot be certain whether E lies inside or outside the triangle until we do the construction.)

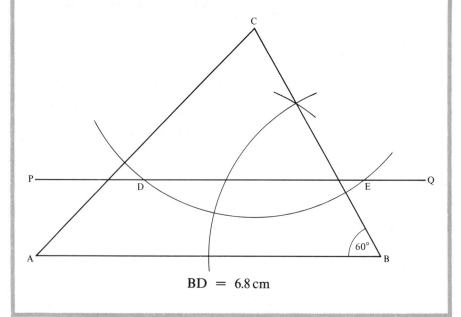

BD = 6.8 cm

14. ABCD is a rectangle with AB = 8 cm and BC = 5 cm. Construct this rectangle and find the point P which is 2 cm from AB and equidistant from AD to BC. Measure PB.

15. Construct a rectangle ABCD with AB = 6.5 cm and AD = 8 cm. Find the point X which is 3 cm from AD and equidistant from AD and DC. Measure DX.

16. Construct a triangle ABC in which AB = 9.5 cm, BC = 8 cm and $A\widehat{B}C = 60°$. Find the point D, on the opposite side of AB from C, that is 7 cm from BC and 4.5 cm from AC. Measure CD.

17. ABC is a triangle in which AB = 12 cm, BC = 9 cm and $A\widehat{B}C = 90°$. Construct this triangle and find a point D that is 4.5 cm from BC and equidistant from A and C. Measure AD.

LOCI INVOLVING REGIONS

A locus is the set of points that satisfies a given condition. If this condition involves an inequality, the set of points is a region rather than a line. For example, the locus of a point P that moves in such a way that it is always 10 cm from a fixed point A, is the *circumference* of a circle, centre A, radius 10 cm, whereas the locus of a point P that moves in such a way that it is always less than 10 cm from a fixed point A, is the region *within the circle* centre A, radius 10 cm.

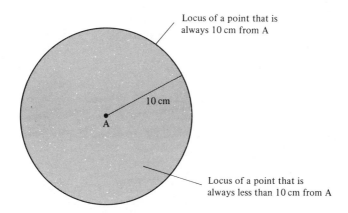

Locus of a point that is always 10 cm from A

10 cm

A

Locus of a point that is always less than 10 cm from A

EXERCISE 13b

A is a fixed point. Show, by shading in a suitable sketch, the locus of P such that AP < 5 cm.

(We begin by drawing the locus of P such that AP = 5 cm. This is a circle, centre A, radius 5 cm.)

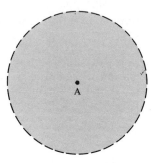

(If P is such that AP < 5 cm, P must be within the circle. P cannot lie on the circumference of the circle. We show this by using a dotted line. If the circle were to be included, it would be drawn as a solid line.)

1. A is a fixed point. Show, by shading in a suitable sketch, the locus of P such that AP ⩽ 6 cm.

2. A is a fixed point. Illustrate the locus of P such that 3 cm < AP < 6 cm.

3. ABC is a triangle. Illustrate the locus of P, within the triangle, such that BP < PC.

4. AB is a fixed line, 10 cm long. Illustrate the locus of P such that P is more than 3 cm from any point on AB but less than 6 cm from any point on AB.

5. AB is a fixed line 10 cm long. Illustrate the locus of a point P such that $A\widehat{P}B < 90°$.

6. AB is a fixed line 10 cm long. Illustrate the locus of a point P that is less than 7 cm from A and more than 8 cm from B.

7. Draw a line AB of length 6 cm. Draw accurately, on the same diagram, the locus of

a) the point P which moves so that AP = PB

b) the point Q such that $\widehat{AQB} = 90°$

c) the point R which moves so that the area of triangle ARB is 6 cm²

Mark on your diagram a point X such that AX < XB, $\widehat{AXB} = 90°$, and the area of triangle AXB is 6 cm². Measure AX.

LOCI IN THREE DIMENSIONS

In three-dimensional work there are two important loci that concern us.

1. If a point moves so that it is always at a fixed distance from a fixed point its locus is a sphere.
   The fixed point is the centre of the sphere and the fixed distance is its radius.

2. The locus of points that are equidistant from two given points A and B is the plane bisecting AB at right angles.

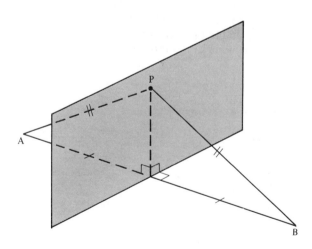

EXERCISE 13e

1. A is a fixed point and a point P moves in space so that AP = 6 cm. Describe the locus of P.

2. A and B are two points 10 cm apart. P is a point in space such that PA = PB. Describe the locus of P.

3. a) A is a fixed point and a point X moves in space so that the length of AX is 5 cm. Describe the locus of X.

b) B is a second fixed point so that AB = 8 cm. A point Y moves in space so that AY = YB. Describe the locus of Y.

c) Describe the set of points of intersection.

4. Describe the locus of a point in space that is always 10 cm from the surface of a sphere of radius 5 cm.

5. Describe the locus of a point in space that is always 5 cm from the surface of a sphere of radius 10 cm. What important difference is there between your answer to this question and your answer to question 4?

6. ABCD is a rectangle with AB = 12 cm and BC = 10 cm.

a) Describe the locus of A as the rectangle is rotated about DC through 360°.

b) Describe the locus of A as the rectangle is rotated about BC through 360°.

c) Describe the locus of A as the rectangle is rotated about BD through 360°.

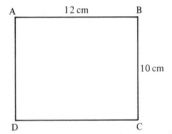

7. ABCD is a rectangle measuring 10 cm by 8 cm.

a) Describe the locus of points 8 cm from the *plane* of the rectangle.

b) Describe the locus of points equidistant from A and D.

c) Describe the intersection of these two loci.

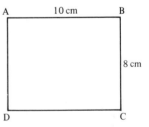

8. A, B and C are three points in space. Describe the locus of points equidistant from A and B and 10 cm from C.

9. Describe the locus of points equidistant from three given points A, B and C.

14 TRANSFORMATIONS

COMMON TRANSFORMATIONS

REFLECTION

A reflection is defined by the mirror line.

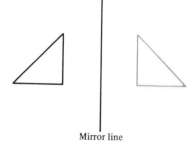

Mirror line

ROTATION

A rotation is defined by the centre and the angle of rotation.

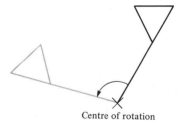

Centre of rotation

TRANSLATION

A translation is defined by a description of the displacement, usually in the form of a vector.

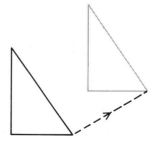

ENLARGEMENT

An enlargement is defined by the centre
of enlargement and the scale factor.

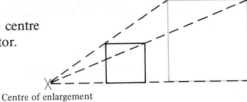

Centre of enlargement

EXERCISE 14a

For each question, draw x and y axes, each for values from -6 to 6.
Use 1 cm to 1 unit.

1. Draw $\triangle ABC$ with A(3, 2), B(5, 2) and C(5, 5). Draw the image of $\triangle ABC$
 a) under a reflection in the line $x = 1$. Label it $\triangle A_1B_1C_1$
 b) under a rotation of 90° clockwise about (0, 0). Label it $\triangle A_2B_2C_2$
 c) under an enlargement, centre (6, 6) and scale factor 3. Label it $\triangle A_3B_3C_3$.

2. Draw $\triangle PQR$ with P(-1, 2), Q(-1, 5) and R(-3, 2).
 Draw the image of $\triangle PQR$
 a) under a reflection in the line $y = x$. Label it $\triangle P_1Q_1R_1$
 b) under a reflection in the y-axis. Label it $\triangle P_2Q_2R_2$
 c) under a reflection in the x-axis. Label it $\triangle P_3Q_3R_3$.
 d) Describe the transformation that maps $\triangle P_2Q_2R_2$ to $\triangle P_3Q_3R_3$.
 e) Describe the transformation that maps $\triangle P_3Q_3R_3$ TO $\triangle P_1Q_1R_1$.

3. Draw rectangle ABCD with A(1, 1), B(4, 1), C(4, 3) and D(1, 3).
 Draw the four images of ABCD under the translations described by the
 vectors
 a) $\begin{pmatrix} -5 \\ 3 \end{pmatrix}$ b) $\begin{pmatrix} -7 \\ -1 \end{pmatrix}$ c) $\begin{pmatrix} -5 \\ -7 \end{pmatrix}$ d) $\begin{pmatrix} 2 \\ -3 \end{pmatrix}$
 Label the images $A_1B_1C_1D_1$, $A_2B_2C_2D_2$, $A_3B_3C_3D_3$ and $A_4B_4C_4D_4$
 respectively.
 e) Describe the transformation that maps $A_4B_4C_4D_4$ to $A_1B_1C_1D_1$.
 f) Describe the transformation that maps $A_2B_2C_2D_2$ to ABCD.

4. Draw $\triangle LMN$ with L(3, 2), M(5, 2) and N(5, 5).

 Draw the image of $\triangle LMN$

 a) under a reflection in the line $y = x$. Label it $\triangle L_1M_1N_1$
 b) under a rotation of 180° about (0, 2). Label it $\triangle L_2M_2N_2$
 c) under a translation described by the vector $\begin{pmatrix} -4 \\ 1 \end{pmatrix}$. Label it $\triangle L_3M_3N_3$.
 d) What is the image of $\triangle LMN$ under a rotation of 360° about O?

5.

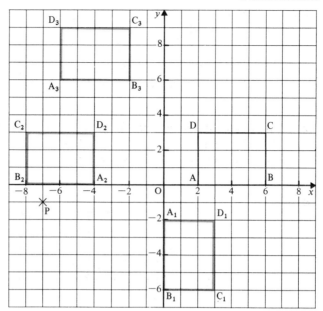

Give the transformation that maps rectangle ABCD to

a) $A_1B_1C_1D_1$ b) $A_2B_2C_2D_2$ c) $A_3B_3C_3D_3$

d) $A_1B_1C_1D_1$ is mapped to $A_3B_3C_3D_3$ by a rotation about $P(-7, -1)$. What is the angle of rotation ?

6.

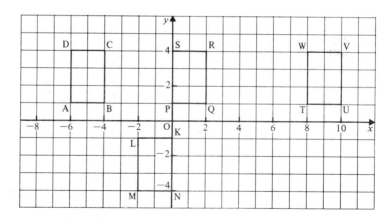

Give the transformation which maps PQRS to

a) ABCD   c) MNKL   e) TUVW
b) KLMN   d) BADC   f) UTWV

Give the transformation which maps ABCD to

g) PQRS   h) QPSR   i) KLMN

7. P is the point $(3, 3)$ and Q is $(1, 2)$. Find the coordinates of the point to which P is mapped under

a) an enlargement, centre Q, scale factor 2

b) an enlargement, centre Q, scale factor -1

c) a clockwise rotation of $90°$ about Q.

8. A translation maps the point $(6, 2)$ to the point $(7, 5)$ and the mirror line of a reflection is the line $x + y = 2$. Find the image of the point $(1, 2)$ under

a) the translation   b) the reflection.

9.

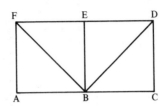

ABEF and BCDE are squares.

a) Under a rotation, $\triangle CDB$ is mapped to $\triangle ABF$. Give the centre of rotation and the angle of rotation.

b) Under a reflection, $\triangle CDB$ is mapped to $\triangle AFB$. Give the mirror line.

c) Under another rotation, square CDEB is mapped to FABE. Give the centre of rotation and the angle of rotation.

10.

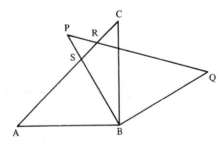

In $\triangle ABC$, $AB = BC$ and $\widehat{ABC} = 90°$. $\triangle ABC$ is mapped to $\triangle PBQ$ by a rotation of $60°$ clockwise about B.

a) Name all the lengths equal to AB and to AC.

b) Through what angle has BC rotated in this transformation? What is the size of $\widehat{CBQ}$?

c) Through what angle has AC rotated? What is the size of $\widehat{ARP}$?

d) Calculate $\widehat{CSB}$.

e) If $AB = 6\,cm$, give the length of CQ.

11.

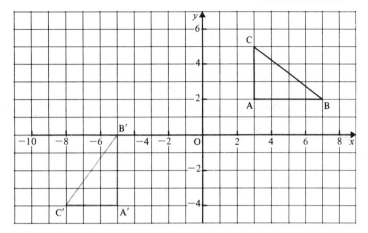

Copy the diagram, using 1 cm to 1 unit. △A′B′C′ is the image of △ABC under a rotation, centre P, where P is the intersection of the perpendicular bisectors of AA′, BB′ and CC′. Use compasses to construct the perpendicular bisectors of CC′ and BB′.

a) Give the coordinates of the centre of rotation and the angle of rotation of this transformation.

b) Explain why the centre of rotation lies on the perpendicular bisectors.

NOTATION

It is useful to have a symbol to describe a transformation.

For example we can use Y to denote a reflection in the *y*-axis. Then the image of a triangle (called P) under the reflection is named Y(P).

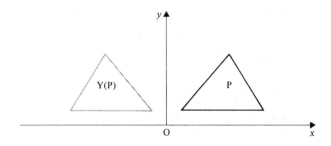

EXERCISE 14b

R₁ is the rotation of 90° anticlockwise about O. The object P is △ABC with vertices A(1, 1), B(4, 1) and C(4, 5). Draw P and R₁(P)

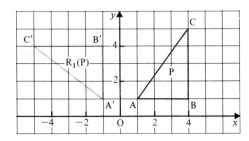

In this exercise,

R₁ is a rotation of 90° anticlockwise about O

R₂ is a rotation of 180° about O

R₃ is a rotation of 90° clockwise about O

X is a reflection in the *x*-axis

Y is a reflection in the *y*-axis

T is a translation defined by the vector $\begin{pmatrix} 1 \\ 3 \end{pmatrix}$.

For each question, draw *x* and *y* axes, each for values from −6 to 6.

1. P, Q and R are the points (2, 1), (5, 1) and (5, 3).
Draw △PQR and label it A.
Draw and label the following images
a) R₁(A)   b) R₂(A)   c) R₃(A)

2. A, B and C are the points (−2, 1), (−4, 1) and (−2, 5).
Draw △ABC and label it Q.
Draw and label the following images
a) X(Q)   b) Y(Q)

3. P is the triangle with vertices ($-5, 0$), ($-3, 0$) and ($-3, 2$).
Q is the triangle with vertices ($1, 0$), ($3, 0$) and ($1, 2$).
R is the triangle with vertices ($-2, -6$), ($0, -6$) and ($0, -4$).
Draw and label   a) T(P)   b) T(Q)   c) T(R)

4. L is a reflection in the line $x + y = 2$ and M is a reflection in the line $y = x + 2$. The object A is the triangle with vertices ($1, 2$), ($4, 2$) and ($4, 4$).
Find   a) L(A)   b) M(A)

5. N is a rotation of 90° anticlockwise about ($1, 1$).
R is a rotation of 90° clockwise about ($0, -1$).
The object B is the triangle with vertices ($-2, 1$), ($-5, 1$) and ($-2, 3$).
Find   a) N(B)   b) R(B)

COMPOUND TRANSFORMATIONS

If we reflect the object P in the *x*-axis and then reflect the resulting image in the *y*-axis, we are carrying out a compound transformation. The letters defined in exercise 14b can be used to describe the final image.

The first image is X(P). The second image is Y(X(P)) but the outer set of brackets is not usually used and it is written YX(P).

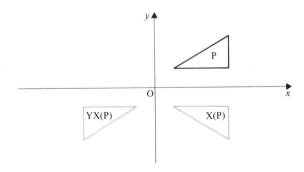

Notice that the letter X, denoting the transformation used first, is nearer to the object P. We work outwards from the bracket containing the object.

EXERCISE 14c

In this exercise,

R_1 is a rotation of 90° anticlockwise about O

R_2 is a rotation of 180° about O

R_3 is a rotation of 90° clockwise about O

X is a reflection in the *x*-axis

Y is a reflection in the *y*-axis

T is a translation defined by the vector $\begin{pmatrix} 4 \\ 3 \end{pmatrix}$.

A, B and C are the points $(4, 1)$, $(6, 1)$ and $(6, 4)$.
Draw △ABC and label it P. Draw and label $R_1(P)$, $XR_1(P)$
and $YR_1(P)$.

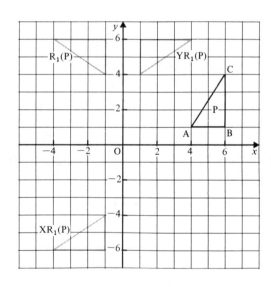

In each of the following questions draw *x* and *y* axes, each for values
from −6 to 6.

1. P, Q and R are the points $(1, 3)$, $(3, 3)$ and $(1, 6)$. Draw △PQR and
label it A. Draw and label

a) $R_3(A)$ b) $XR_3(A)$ c) $X(A)$ d) $R_3X(A)$

e) Describe the single transformation that will map A to $R_3X(A)$.

2. L, M and N are the points $(-2, 1)$, $(-4, 1)$ and $(-2, 5)$. Draw $\triangle$LMN and label it P. Draw and label

a) $R_1(P)$ b) $R_2R_1(P)$ c) $R_3R_1(P)$

d) What is the single transformation that will map P to $R_2R_1(P)$?

e) What is the single transformation that will map $R_2R_1(P)$ to $R_1(P)$?

3. A, B and C are the points $(-5, 2)$, $(-2, 2)$ and $(-5, 4)$. Draw $\triangle$ABC and label it P. Draw and label

a) $X(P)$ b) $YX(P)$ c) $R_2(P)$ d) $Y(P)$ e) $XY(P)$

f) Is $YX(P)$ the same triangle as $XY(P)$?

g) Is $R_2(P)$ the same triangle as $XY(P)$?

h) What single transformation is equivalent to a reflection in the x-axis followed by a reflection in the y-axis?

4. L, M and N are the points $(-3, 0)$, $(-1, 0)$ and $(-1, 3)$. Draw $\triangle$LMN and label it Q. Find

a) $T(Q)$ b) $XT(Q)$ c) $X(Q)$ d) $TX(Q)$

e) Describe the single transformation that will map $X(Q)$ to $XT(Q)$.

f) Describe the single transformation that will map $XT(Q)$ to $TX(Q)$.

EQUIVALENT SINGLE TRANSFORMATIONS

We have seen that if we reflect an object P in the x-axis and then reflect the image $X(P)$ in the y-axis we get the same final image as if we had rotated P through $180°$ about O. $YX(P)$ is the same as $R_2(P)$ and the effect of YX is the same as the effect of R_2.

We can write $YX(P) = R_2(P)$ referring to the images
and $YX = R_2$ referring to the transformations.

In the following exercise notice that $X^2 = XX$, i.e. the transformation X is used twice in succession.

EXERCISE 14d

In each question draw x and y axes, each for values from -6 to 6. Use 1 cm for 1 unit.

1. A is a reflection in the line $x = -1$ and B a reflection in the line $y = 2$. Label as Z the triangle PQR where P is the point $(1, 4)$, $Q(4, 6)$ and $R(1, 6)$.

a) Find $A(Z)$, $B(Z)$, $AB(Z)$ and $BA(Z)$.

b) Describe the single transformations given by AB and BA. Is AB equal to BA?

c) Find $A^2(Z)$ and $B^2(Z)$.

2. T is a reflection in the line $x = 1$.
U is a reflection in the line $y = 2$.
V is a rotation of $180°$ about the point $(1, 2)$.
Label with A the triangle PQR where P is the point $(1, 1)$, Q is $(3, 1)$ and R is $(3, -2)$.

a) Draw T(A), U(A), TU(A) and UT(A)

b) Are TU and UT the same transformation?

c) Is it true that V = TU?

3. R_1 is a rotation of $90°$ anticlockwise about O.
R_2 is a rotation of $180°$ about O.
R_3 is a rotation of $90°$ clockwise about O.
Label with P the triangle ABC where A is the point $(1, 2)$, B is $(4, 2)$ and C is $(1, 4)$.

a) Draw $R_1(P)$, $R_1{}^2(P)$, $R_2(P)$, $R_2R_1(P)$ and $R_3(P)$.
Complete the following statement
$$R_1{}^2 = \qquad \text{and } R_2R_1 =$$

b) Draw whatever images are needed and complete the following statements
$$R_3{}^2 = \qquad R_2R_3 = \qquad \text{and } R_3R_2 =$$

THE IDENTITY TRANSFORMATION

If an object is rotated through $360°$ or translated using the vector $\begin{pmatrix} 0 \\ 0 \end{pmatrix}$, the final image turns out to be the same as the original object. We are back where we started and might as well not have performed a transformation at all. This operation is called the *identity transformation* and is usually named I.

EXERCISE 14e

In this exercise

R_1 is a rotation of $90°$ anticlockwise about O

R_2 is a rotation of $180°$ about O

R_3 is a rotation of $90°$ clockwise about O

A is a reflection in the x-axis

B is a reflection in the y-axis

C is a reflection in the line $y = x$

D is a reflection in the line $y = -x$

I is the identity transformation.

P is the triangle with vertices (2, 1), (5, 1) and (2, 3).
Find B(P) and B²(P). Name the single transformation which is
equal to B².

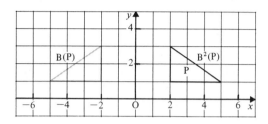

$$B^2(P) = P$$

$$\therefore \qquad B^2 = I$$

For each question draw x and y axes, each for values from −6 to 6.
Use 1 cm to 1 unit.

1. P is the triangle with vertices (2, 1), (5, 1) and (5, 5).
   a) Find $R_1(P)$, $R_2(P)$, $R_3(P)$, $R_1R_3(P)$ and $R_3R_1(P)$.
   Name the single transformation which is equal to R_1R_3 and to R_3R_1.
   b) Complete the following statements with a single letter.
   i) $R_2{}^2 =$ ii) $R_2R_3 =$ iii) $R_1R_2 =$

2. Q is the triangle with vertices (−2, 1), (−5, 1) and (−4, 5).
   a) Find A(Q), B(Q), AB(Q), R_2(Q) and B²(Q).
   b) Complete the following statements with a single letter.
   i) $B^2 =$ ii) AB =

3. N is the triangle with vertices (1, 3), (1, 6) and (5, 6).
   a) Find C(N), DC(N), C²(N), AC(N), BC(N) and IC(N).
   b) Complete the following statements with a single letter.
   i) DC =   ii) C² =   iii) AC =
   iv) BC =   v) IC =

4. M is the triangle with vertices $(3, 2)$, $(5, 2)$ and $(5, 6)$.
a) Find $R_1(M)$, $A(M)$, $A^2(M)$, $AR_1(M)$ and $R_1A(M)$.
b) Complete the following statements with a single letter.
i) $A^2 =$ ii) $AR_1 =$ iii) $R_1A =$
c) Is the statement $AR_1 = R_1A$ true or false?

5. L is the triangle with vertices $(3, 1)$, $(4, 4)$ and $(1, 4)$.
a) Find $I(L)$, $AI(L)$, $BI(L)$, $IA(L)$ and $IB(L)$.
b) Simplify AI, BI, IA and IB.

6. P is the rectangle with vertices $(-2, -2)$, $(-5, -2)$, $(-5, -4)$ and $(-2, -4)$.
a) Find $I(P)$, $CI(P)$, $DI(P)$ and $D^2(P)$.
b) Simplify CI, DI and D^2.

7. Q is the rhombus with vertices $(-3, -3)$, $(-4, -1)$, $(-3, -1)$ and $(-2, -1)$.
a) Find $R_1(Q)$, $R_2R_1(Q)$ and $R_1R_2R_1(Q)$.
b) Simplify R_2R_1 and $R_1R_2R_1$.
c) Is it true that $R_1R_3 = R_3R_1 = R_2^2$?

VECTORS

Vectors are quantities with both magnitude (i.e. size) and direction.
Examples of vector quantities are velocity and force.
Most vectors are *free vectors* and are not fixed in position.

In the diagram, for example, both $\overrightarrow{AD}$ and $\overrightarrow{BC}$ can be labelled **a** and represent the same vector.

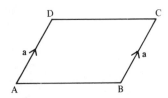

We use these vectors when working with *displacements,* i.e. position changes, as in question 3 in the first exercise in this chapter.

POSITION VECTORS

Position vectors are used to describe the positions of points. Therefore, on a set of axes, the *position vector* of a point A is the vector from a given fixed point, usually the origin, to A.

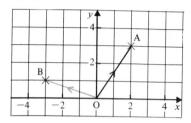

Points A and B have coordinates (2, 3) and (−3, 1). The position vector of A relative to the origin is $\begin{pmatrix} 2 \\ 3 \end{pmatrix}$ and the position vector of B is $\begin{pmatrix} -3 \\ 1 \end{pmatrix}$.

MATRIX TRANSFORMATIONS

If we use a transformation matrix to find the image of point A under a transformation then it is really the image of the position vector that we are finding.

If the transformation matrix is $\begin{pmatrix} 1 & 3 \\ 1 & 0 \end{pmatrix}$ then $\begin{pmatrix} 1 & 3 \\ 1 & 0 \end{pmatrix}\overset{A}{\begin{pmatrix} 2 \\ 3 \end{pmatrix}} = \overset{A'}{\begin{pmatrix} 11 \\ 2 \end{pmatrix}}$

The position vector of A′ is $\begin{pmatrix} 11 \\ 2 \end{pmatrix}$ and A′ is the point (11, 2)

We can save time when finding the images of several points, by lining up their position vectors side by side

$$\begin{pmatrix} 1 & 3 \\ 1 & 0 \end{pmatrix}\overset{A \quad B \quad C}{\begin{pmatrix} 2 & -3 & 1 \\ 3 & 1 & 4 \end{pmatrix}} = \overset{A' \quad B' \quad C'}{\begin{pmatrix} 11 & 0 & 13 \\ 2 & -3 & 1 \end{pmatrix}}$$

Then we can see that the position vectors of the images of B′ and C′ are $\begin{pmatrix} 0 \\ -3 \end{pmatrix}$ and $\begin{pmatrix} 13 \\ 1 \end{pmatrix}$ respectively, so B′ is the point(0, −3) and C′ is the point (13, 1).

EXERCISE 14f

Draw △ABC with A(2, 1), B(5, 1) and C(4, 4). Find the image of △ABC under the transformation defined by the matrix $\begin{pmatrix} -1 & 0 \\ 0 & 1 \end{pmatrix}$. Describe the transformation.

$$\begin{pmatrix} -1 & 0 \\ 0 & 1 \end{pmatrix} \overset{\begin{matrix} A & B & C \end{matrix}}{\begin{pmatrix} 2 & 5 & 4 \\ 1 & 1 & 4 \end{pmatrix}} = \overset{\begin{matrix} A' & B' & C' \end{matrix}}{\begin{pmatrix} -2 & -5 & -4 \\ 1 & 1 & 4 \end{pmatrix}}$$

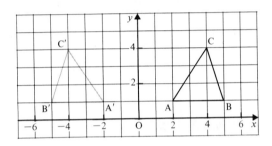

The transformation is a reflection in the *y*-axis.

For each of the questions from 1 to 3 draw *x* and *y* axes for values in the given ranges and draw △ABC with A(1, 0), B(4, 0) and C(4, 2).

1. Use $-4 \leqslant x \leqslant 4$ and $-4 \leqslant y \leqslant 4$.
Find the images of △ABC under the transformations defined by the matrices

a) $\begin{pmatrix} 0 & -1 \\ 1 & 0 \end{pmatrix}$ b) $\begin{pmatrix} -1 & 0 \\ 0 & -1 \end{pmatrix}$ c) $\begin{pmatrix} 0 & 1 \\ -1 & 0 \end{pmatrix}$ d) $\begin{pmatrix} 1 & 0 \\ 0 & -1 \end{pmatrix}$

Describe each of the four transformations.

2. Use $-4 \leqslant x \leqslant$ and $-4 \leqslant y \leqslant 4$.
Find the images of △ABC under the transformations defined by the matrices

a) $\begin{pmatrix} 0 & 1 \\ 1 & 0 \end{pmatrix}$ b) $\begin{pmatrix} -1 & 0 \\ 0 & 1 \end{pmatrix}$ c) $\begin{pmatrix} 0 & -1 \\ -1 & 0 \end{pmatrix}$

Describe each of the three transformations.

3. Use $-12 \leqslant x \leqslant 8$ and $-6 \leqslant y \leqslant 4$.

Find the images of $\triangle ABC$ under the transformations defined by the matrices

a) $\begin{pmatrix} 2 & 0 \\ 0 & 2 \end{pmatrix}$ b) $\begin{pmatrix} -3 & 0 \\ 0 & -3 \end{pmatrix}$ c) $\begin{pmatrix} -\frac{1}{2} & 0 \\ 0 & -\frac{1}{2} \end{pmatrix}$

Describe each transformation.

4. Draw x and y axes for $-6 \leqslant x \leqslant 6$ and $0 \leqslant y \leqslant 4$.

Draw rectangle PQRS with P(1, 0), Q(3, 0), R(3, 3) and S(1, 3).

Find the images of PQRS under the transformations defined by the matrices

a) $\begin{pmatrix} 1 & 1 \\ 0 & 1 \end{pmatrix}$ b) $\begin{pmatrix} 1 & -2 \\ 0 & 1 \end{pmatrix}$

5. Draw x and y axes each for values from 0 to 6. Draw the rectangle ABCD where A is (0, 1), B is (3, 1), C is (3, 3) and D is (0, 3).

Find the images of ABCD under the transformations defined by the matrices

a) $\begin{pmatrix} 1 & 0 \\ 0 & 2 \end{pmatrix}$ b) $\begin{pmatrix} 3 & 0 \\ 0 & 1 \end{pmatrix}$

6. Draw x and y axes each for values from -6 to 6. Use the square OABC as the object, where O is (0, 0), A is (1, 0), B is (1, 1) and C is (0, 1).

Make up two enlargement matrices and test them on the square OABC. Describe the enlargement in each case.

COMMON TRANSFORMATION MATRICES

It is useful to know the following facts.

a) An enlargement matrix is of the form $\begin{pmatrix} k & 0 \\ 0 & k \end{pmatrix}$

b) Reflection in the x or y-axis or lines $y = \pm x$ and rotations of multiples of $90°$ about the origin are produced by matrices with zeros in one diagonal and 1 or -1 elsewhere,

e.g. $\begin{pmatrix} 0 & -1 \\ 1 & 0 \end{pmatrix}$ gives a rotation of $180°$ about O.

c) There are no matrices that produce reflections in lines other than lines through the origin, nor are there any matrices that produce rotation about points other than the origin.

d) The unit matrix $\begin{pmatrix} 1 & 0 \\ 0 & 1 \end{pmatrix}$ maps an object to itself. This matrix gives the identity transformation (for instance, a rotation of $360°$ about O).

e) A translation is *not* produced by a matrix but is defined by a vector.

INVERSE MATRICES AND DETERMINANTS

When a matrix is multiplied by its inverse, the result is the *identity matrix*.

To find the inverse of $\mathbf{M} = \begin{pmatrix} 2 & 1 \\ 4 & 3 \end{pmatrix}$, we first interchange the entries in

the leading diagonal, $\begin{pmatrix} 3 & \\ & 2 \end{pmatrix}$, and change the sign of the entries in the

other diagonal, $\begin{pmatrix} & -1 \\ -4 & \end{pmatrix}$.

Then we divide the resulting matrix, $\begin{pmatrix} 3 & -1 \\ -4 & 2 \end{pmatrix}$, by the *determinant* of $\mathbf{M}$.

The determinant of a matrix can be found from a formula,

i.e. the determinant of $\begin{pmatrix} a & b \\ c & d \end{pmatrix}$ is $ad - bc$

In the example given in the previous section the determinant of $\mathbf{M}$ is

$$2 \times 3 - 1 \times 4 \qquad \text{i.e. } 2$$

Alternatively the determinant may be found by multiplying the original

matrix $\begin{pmatrix} 2 & 1 \\ 4 & 3 \end{pmatrix}$ by the attempt at the inverse.

In this case, $\begin{pmatrix} 3 & -1 \\ -4 & 2 \end{pmatrix}\begin{pmatrix} 2 & 1 \\ 4 & 3 \end{pmatrix} = \begin{pmatrix} 2 & 0 \\ 0 & 2 \end{pmatrix}$

and we can see that the determinant is 2.

The inverse of $\begin{pmatrix} 2 & 1 \\ 4 & 3 \end{pmatrix}$ is $\frac{1}{2}\begin{pmatrix} 3 & -1 \\ -4 & 2 \end{pmatrix} = \begin{pmatrix} \frac{3}{2} & -\frac{1}{2} \\ -2 & 1 \end{pmatrix}$

Similarly, for $\begin{pmatrix} 5 & -6 \\ -3 & 4 \end{pmatrix}$, the determinant given by the formula is

$$5 \times 4 - (-6) \times (-3) \qquad \text{i.e. } 2$$

Hence the inverse of $\begin{pmatrix} 5 & -6 \\ -3 & 4 \end{pmatrix}$ is $\frac{1}{2}\begin{pmatrix} 4 & 6 \\ 3 & 5 \end{pmatrix} = \begin{pmatrix} 2 & 3 \\ \frac{3}{2} & \frac{5}{2} \end{pmatrix}$

EXERCISE 14g

Find the determinant of the matrix $\begin{pmatrix} 3 & 4 \\ 1 & 2 \end{pmatrix}$

First method   (Using the formula)

The determinant of $\begin{pmatrix} 3 & 4 \\ 1 & 2 \end{pmatrix}$ is $(3 \times 2) - (4 \times 1)$

$$= 6 - 4$$

$$= 2$$

Second method   (Start by trying to find the inverse)

Try $\begin{pmatrix} 2 & -4 \\ -1 & 3 \end{pmatrix}$ as the inverse

$$\begin{pmatrix} 3 & 4 \\ 1 & 2 \end{pmatrix} \begin{pmatrix} 2 & -4 \\ -1 & 3 \end{pmatrix} = \begin{pmatrix} 2 & 0 \\ 0 & 2 \end{pmatrix}$$

$\therefore$ the determinant is 2.

Find the determinants of the following matrices.

1. $\begin{pmatrix} 3 & 2 \\ 1 & 2 \end{pmatrix}$ **4.** $\begin{pmatrix} 4 & -2 \\ 1 & 1 \end{pmatrix}$ **7.** $\begin{pmatrix} 5 & -1 \\ -3 & 2 \end{pmatrix}$

2. $\begin{pmatrix} 9 & 2 \\ 4 & 1 \end{pmatrix}$ **5.** $\begin{pmatrix} 4 & -2 \\ 1 & -1 \end{pmatrix}$ **8.** $\begin{pmatrix} -1 & 0 \\ 0 & -1 \end{pmatrix}$

3. $\begin{pmatrix} 4 & 2 \\ 3 & 1 \end{pmatrix}$ **6.** $\begin{pmatrix} -2 & 4 \\ 1 & -3 \end{pmatrix}$ **9.** $\begin{pmatrix} 2 & 6 \\ 1 & 3 \end{pmatrix}$

In each question from 10 to 15, draw x and y axes each for values from -2 to 6. Draw the object OABC where A is $(2, 0)$, B is $(2, 1)$ and C is $(0, 1)$. Find the image of OABC under the transformation given by the matrix and answer the following questions.

a) Find the areas of OABC and the image OA'B'C'.

b) Find the value of $\dfrac{\text{area of OA'B'C'}}{\text{area of OABC}}$

c) Find the determinant of the transformation matrix. What do you notice?

Carry out instructions (a) to (c) for each of the following matrices.

10. $\begin{pmatrix} -1 & 0 \\ 0 & -1 \end{pmatrix}$ **12.** $\begin{pmatrix} 3 & 0 \\ 0 & 3 \end{pmatrix}$ **14.** $\begin{pmatrix} 0 & 0 \\ 0 & 0 \end{pmatrix}$

11. $\begin{pmatrix} 1 & 2 \\ 0 & 1 \end{pmatrix}$ **13.** $\begin{pmatrix} \frac{1}{2} & 0 \\ 0 & \frac{1}{2} \end{pmatrix}$ **15.** $\begin{pmatrix} 2 & 2 \\ 0 & 3 \end{pmatrix}$

AREA SCALE FACTOR

The fraction $\dfrac{\text{area of the image}}{\text{area of the object}}$ is the *area scale factor,* and we see from the results of the previous exercise that it is given by the determinant of the transformation matrix.

If we know the area of the object, the determinant can be used to calculate the area of the image.

UNIT VECTORS AND THE UNIT SQUARE

In question 6 of exercise 14f the object was taken as the square OABC where A is the point (1, 0) and C is (0, 1).

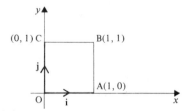

This is a particularly easy object to use, as the position vectors

of A and C are $\begin{pmatrix} 1 \\ 0 \end{pmatrix}$ and $\begin{pmatrix} 0 \\ 1 \end{pmatrix}$ and, lined up together, they form

the unit matrix $\begin{pmatrix} 1 & 0 \\ 0 & 1 \end{pmatrix}$

The vectors $\begin{pmatrix} 1 \\ 0 \end{pmatrix}$ and $\begin{pmatrix} 0 \\ 1 \end{pmatrix}$ are called the *base vectors* for the coordinates.

Often O, A′ and C′ are enough to identify the image of the unit square but, if the shape of the image is still not clear, we can also find the image of B(1, 1).

EXERCISE 14h

Using the unit square as the object, find the image under the transformation defined by the matrix given in each question from 1 to 6. If it is possible to do so, identify the transformation, describing it fully.

1. $\begin{pmatrix} 0 & -1 \\ 1 & 0 \end{pmatrix}$ **3.** $\begin{pmatrix} -1 & 0 \\ 0 & -1 \end{pmatrix}$ **5.** $\begin{pmatrix} 1 & 2 \\ 0 & 1 \end{pmatrix}$

2. $\begin{pmatrix} 1 & -2 \\ 0 & 1 \end{pmatrix}$ **4.** $\begin{pmatrix} 2\frac{1}{2} & 0 \\ 0 & 2\frac{1}{2} \end{pmatrix}$ **6.** $\begin{pmatrix} 3 & 12 \\ 1 & 4 \end{pmatrix}$

7. Find the area scale factors of the transformations given by the matrices in questions 1 to 6.

8. Given the matrix $\begin{pmatrix} 4 & -8 \\ -1 & 2 \end{pmatrix}$

     a) find the image of the unit square under the transformation given by the matrix

     b) find the determinant of the matrix and the area scale factor of the transformation

     c) find the area of the image.

9. A rectangle ABCD has area 6 square units. It is transformed using the matrix $\begin{pmatrix} 3 & 1 \\ 1 & 2 \end{pmatrix}$

     a) Find the determinant of the matrix and the area scale factor of the transformation.

     b) Find the area of the image of ABCD.

10. A, B, C and D are the points (0, 2), (0, 4), (3, 4) and (3, 2) respectively.

     a) Find the area of rectangle ABCD.

     b) ABCD is transformed using the matrix $\begin{pmatrix} 4 & 5 \\ 1 & 2 \end{pmatrix}$

     Calculate the area of the image of ABCD (do *not* find and draw the image).

INVERSE TRANSFORMATIONS

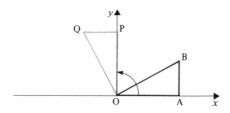

A rotation of 90° anticlockwise about O maps △OAB to △OPQ.
The *inverse* transformation is a rotation of 90° clockwise about O.
It maps △OPQ back to △OAB.
Some transformations, such as reflections, are their own inverses.
Some transformations have no inverses as the next exercise will show.

EXERCISE 14i

Describe the inverses of the following transformations.

1. A rotation of 60° clockwise about the point (1, 1).

2. A reflection in the *x*-axis.

3. An enlargement, centre (0, 0) and scale factor 2.

4. A translation defined by the vector $\begin{pmatrix} 2 \\ 3 \end{pmatrix}$.

5. A rotation of 180° about O.

For each question from 6 to 9, draw *x* and *y* axes each for values
from −4 to 4. Use the unit square OABC as the object. A is the point (1, 0),
B is (1, 1) and C is (0, 1).

6. Given the matrix $\begin{pmatrix} 0 & 1 \\ -1 & 0 \end{pmatrix}$

a) Find the image of the unit square under the transformation given by the matrix.

b) Describe the transformation.

c) Describe the inverse transformation.

d) Find the inverse of the matrix $\begin{pmatrix} 0 & 1 \\ -1 & 0 \end{pmatrix}$

e) Transform OA′B′C′ using the inverse matrix. What happens ?

f) Describe the transformation given by the inverse matrix in part (e).

g) Are the transformations described in parts (c) and (f) the same ?
   What is the matrix of the inverse transformation ?

7. Repeat question 6 with the matrix $\begin{pmatrix} 2 & 0 \\ 0 & 2 \end{pmatrix}$

8. Repeat question 6 with the matrix $\begin{pmatrix} 0 & -1 \\ 1 & 0 \end{pmatrix}$

9. Repeat question 6 parts (a) to (d) with matrices

   i) $\begin{pmatrix} 0 & -1 \\ -1 & 0 \end{pmatrix}$ ii) $\begin{pmatrix} -1 & 0 \\ 0 & 1 \end{pmatrix}$

   Comment on the result.

10. Draw x and y axes, each for values from -6 to 6. The object is the unit square.
   Repeat question 6 parts (a) and (b) with the matrices

   i) $\begin{pmatrix} 3 & 6 \\ 1 & 2 \end{pmatrix}$ ii) $\begin{pmatrix} 3 & 2 \\ 6 & 4 \end{pmatrix}$

   What goes wrong in each case when you try to describe the inverse transformation?

FINDING A TRANSFORMATION MATRIX

EXERCISE 14k

A is the point $(2, 1)$, B is $(3, 4)$, P is $(5, 3)$ and Q is $(10, 12)$.
Find the matrix of the transformation under which
AB is mapped to PQ.

Let the transformation matrix be $\begin{pmatrix} a & b \\ c & d \end{pmatrix}$

then

$$\begin{pmatrix} a & b \\ c & d \end{pmatrix} \overset{A\ B}{\begin{pmatrix} 2 & 3 \\ 1 & 4 \end{pmatrix}} = \overset{P\ \ Q}{\begin{pmatrix} 5 & 10 \\ 3 & 12 \end{pmatrix}}$$

$$\begin{pmatrix} 2a + b & 3a + 4b \\ 2c + d & 3c + 4d \end{pmatrix} = \begin{pmatrix} 5 & 10 \\ 3 & 12 \end{pmatrix}$$

Comparing entries in the first row gives

$$2a + b = 5 \qquad (1)$$
$$3a + 4b = 10 \qquad (2)$$

$(1) \times 4$

$$8a + 4b = 20 \qquad (3)$$
$$3a + 4b = 10 \qquad (2)$$

$$(3) - (2) \qquad\qquad 5a = 10$$
$$a = 2$$

In (1) $\qquad\qquad b = 1$

Comparing entries in the second row gives

$$2c + d \; = 3 \qquad (4)$$
$$3c + 4d = 12 \qquad (5)$$

$(4) \times 4 \qquad\qquad 8c + 4d = 12 \qquad (6)$
$$3c + 4d = 12 \qquad (5)$$

$(6) - (5) \qquad\qquad 5c = 0$
$$c = 0$$

In (5) $\qquad\qquad d = 3$

$\therefore$ the matrix is $\begin{pmatrix} 2 & 1 \\ 0 & 3 \end{pmatrix}$

In each question from 1 to 6 find the matrix of the transformation which maps AB to PQ.

1. A is $(1, 1)$, B is $(2, 1)$, P is $(1, 2)$ and Q is $(2, 2)$

2. A is $(1, 1)$, B is $(3, 1)$, P is $(3, 2)$ and Q is $(5, 6)$

3. A is $(4, 2)$, B is $(1, 1)$, P is $(12, 2)$ and Q is $(3, 2)$

4. A is $(2, 1)$, B is $(1, 1)$, P is $(1, 5)$ and Q is $(0, 5)$

5. A is $(2, 2)$, B is $(1, 3)$, P is $(-6, 6)$ and Q is $(-11, 9)$

6. A is $(1, 2)$, B is $(1, -2)$, P is $(1, 0)$ and Q is $(17, -4)$

7.

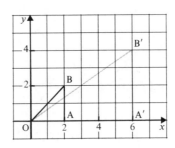

A transformation maps $\triangle OAB$ to $\triangle OA'B'$. Find the matrix that defines this transformation. (Use AB and its image A'B'.)

8.

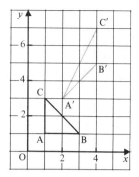

A transformation maps △ABC to △A′B′C′. Find the matrix that defines this transformation. (Use two points to find the matrix. The third point may be used for a check.)

Find the matrix that defines reflection in the *y*-axis.

(We may choose our own object so we use the unit square.)

Let the matrix be $\begin{pmatrix} a & b \\ c & d \end{pmatrix}$

$$\begin{array}{c} \phantom{\begin{pmatrix}a&b\\c&d\end{pmatrix}}\ \ \overset{\text{A}\ \ \text{B}}{} \ \ \ \overset{\text{A}'\ \ \text{B}'}{} \\ \begin{pmatrix} a & b \\ c & d \end{pmatrix} \begin{pmatrix} 1 & 0 \\ 0 & 1 \end{pmatrix} = \begin{pmatrix} -1 & 0 \\ 0 & 1 \end{pmatrix} \end{array}$$

$\therefore \qquad \begin{pmatrix} a & b \\ c & d \end{pmatrix} = \begin{pmatrix} -1 & 0 \\ 0 & 1 \end{pmatrix}$

i.e. the transformation matrix is $\begin{pmatrix} -1 & 0 \\ 0 & 1 \end{pmatrix}$

In each question from 9 to 12, find the matrix that defines the transformation.

9. Reflection in the x-axis.

10. Rotation of 90° clockwise about O.

11. Rotation of 90° anticlockwise about O.

12. Reflection in the line $y = x$.

MIXED QUESTIONS

EXERCISE 14I

1. Find the value of a for which the matrix $\begin{pmatrix} a-2 & 0 \\ 0 & 3 \end{pmatrix}$

a) has no inverse

b) represents an enlargement. State the scale factor of this enlargement.

2. Under a certain transformation, (x', y') is the image of (x, y) and

$$\begin{pmatrix} x' \\ y' \end{pmatrix} = \begin{pmatrix} 3 & 1 \\ 1 & 1 \end{pmatrix} \begin{pmatrix} x \\ y \end{pmatrix}$$

a) Find the coordinates of the image of the point $(2, 3)$.

b) Find the coordinates of the image of the point $(-1, 2)$.

c) Find the coordinates of the point of which $(7, 3)$ is the image.

3. A transformation T is defined by the matrix $\begin{pmatrix} 4 & 2 \\ 7 & 4 \end{pmatrix}$

a) Find the inverse of the matrix.

b) Given that T maps the point A to the point $(8, 15)$, find the coordinates of A.

4. Find the matrix product $\begin{pmatrix} 0 & 1 \\ 1 & 0 \end{pmatrix} \begin{pmatrix} x \\ y \end{pmatrix}$ and describe the transformation defined by the matrix $\begin{pmatrix} 0 & 1 \\ 1 & 0 \end{pmatrix}$

<u>5.</u> A transformation is defined by the matrix $\begin{pmatrix} 1 & 0 \\ 3 & 1 \end{pmatrix}$

a) Express as a single matrix $\begin{pmatrix} 1 & 0 \\ 3 & 1 \end{pmatrix} \begin{pmatrix} x \\ y \end{pmatrix}$

b) Find the coordinates of the image of the point (p, p) under the transformation.

c) The transformation maps the line $y = x$ to the line $y = mx$. Find the value of m.

EXERCISE 14m

In this exercise, several alternative answers are given. Write down the letter that corresponds to the correct answer.

1. The determinant of the matrix $\begin{pmatrix} 4 & 6 \\ 2 & 4 \end{pmatrix}$ is

    A 16   **B** 4   **C** −16   **D** $\frac{1}{4}$

2. The inverse of the matrix $\begin{pmatrix} 3 & 1 \\ 4 & 2 \end{pmatrix}$ is

    A $\begin{pmatrix} 1 & -\frac{1}{2} \\ -2 & 1\frac{1}{2} \end{pmatrix}$ **B** $\begin{pmatrix} 2 & -1 \\ -4 & 3 \end{pmatrix}$ **C** $\begin{pmatrix} 1\frac{1}{2} & \frac{1}{2} \\ 2 & 1 \end{pmatrix}$ **D** $\begin{pmatrix} \frac{1}{5} & \frac{1}{10} \\ -\frac{2}{5} & \frac{3}{10} \end{pmatrix}$

3. The inverse of rotation of 270° clockwise about O is

    A rotation of 90° clockwise about O

    B rotation of 270° anticlockwise about (1, 0)

    C rotation of 90° anticlockwise about O

    D rotation of 270° clockwise about O

4. The area scale factor of the transformation defined by the matrix $\begin{pmatrix} 3 & -1 \\ 2 & 1 \end{pmatrix}$ is

    A 1   **B** $\frac{1}{5}$ **C** 2   **D** 5

5.

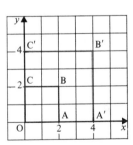

The matrix which maps OABC to OA′B′C′ is

    A $\begin{pmatrix} 1 & 0 \\ 0 & 1 \end{pmatrix}$ **B** $\begin{pmatrix} 2 & 0 \\ 0 & 2 \end{pmatrix}$ **C** $\begin{pmatrix} 0 & 2 \\ 2 & 0 \end{pmatrix}$ **D** $\begin{pmatrix} \frac{1}{2} & 0 \\ 0 & \frac{1}{2} \end{pmatrix}$

15 QUADRATIC EQUATIONS

COMPLETING THE SQUARE

The numbers 9 and 25 can be written as 3^2 and 5^2, whereas 7 and 11 cannot be written as the square of another exact number. Because 9 and 25 can be written in this way they are called *perfect squares*.

Other examples are $\frac{9}{4}$ which is $\left(\frac{3}{2}\right)^2$ and $\frac{121}{169}$ which is $\left(\frac{11}{13}\right)^2$.

In a similar way, because we can write $x^2 + 2x + 1$ as $(x + 1)^2$ and $4x^2 + 12x + 9$ as $(2x + 3)^2$, we say that $x^2 + 2x + 1$ and $4x^2 + 12x + 9$ are perfect squares.

EXERCISE 15a

Express $x^2 + 12x + 36$ in the form $(x + a)^2$

$$x^2 + 12x + 36 = (x + 6)(x + 6)$$
$$= (x + 6)^2$$

Express each of the following expressions in the form $(x + a)^2$.

1. $x^2 + 6x + 9$

2. $a^2 + 4a + 4$

3. $p^2 - 10p + 25$

4. $s^2 - 12s + 36$

5. $x^2 - 5x + \frac{25}{4}$

6. $b^2 + 3b + \frac{9}{4}$

7. $x^2 + 9x + \frac{81}{4}$

8. $a^2 - a + \frac{1}{4}$

9. $x^2 - \frac{1}{2}x + \frac{1}{16}$

10. $x^2 + 8x + 16$

11. $x^2 + x + \frac{1}{4}$

12. $x^2 + \frac{2}{3}x + \frac{1}{9}$

13. $p^2 + 18p + 81$ **15.** $t^2 - \frac{3}{2}t + \frac{9}{16}$ **17.** $x^2 - 2cx + c^2$

14. $a^2 - \frac{4}{5}a + \frac{4}{25}$ **16.** $x^2 + 2bx + b^2$ **18.** $x^2 + \frac{b}{a}x + \frac{b^2}{4a^2}$

Express $4x^2 + 12x + 9$ in the form $(ax + b)^2$

$$4x^2 + 12x + 9 = (2x + 3)(2x + 3)$$
$$= (2x + 3)^2$$

Express the following expressions in the form $(ax + b)^2$.

19. $9x^2 + 6x + 1$ **22.** $9x^2 - 24x + 16$ **25.** $9x^2 - 6x + 1$

20. $4x^2 - 12x + 9$ **23.** $4x^2 - 4x + 1$ **26.** $4x^2 + 2x + \frac{1}{4}$

21. $100x^2 - 60x + 9$ **24.** $25x^2 + 20x + 4$ **27.** $\frac{9x^2}{4} + 2x + \frac{4}{9}$

FORMING A PERFECT SQUARE

To make $x^2 + 6x$ into a perfect square we must add 9 to it.

Then $x^2 + 6x + 9 = (x + 3)^2$

and to make $x^2 - 3x$ into a perfect square we must add $\frac{9}{4}$ to it.

Then $x^2 - 3x + \frac{9}{4} = \left(x - \frac{3}{2}\right)^2$

More generally, to make $x^2 + px$ into a perfect square we take half the coefficient of x $\left(\text{i.e.} \frac{p}{2}\right)$, square it $\left(\text{i.e.} \frac{p^2}{4}\right)$ and add this to $x^2 + px$.

Then $x^2 + px + \frac{p^2}{4}$ is a perfect square and can be written $\left(x + \frac{p}{2}\right)^2$

EXERCISE 15b

What must be added to $x^2 + 6x$ to make it into a perfect square ?

The coefficient of x is 6

Half of this is 3

The square of 3 is 9

9 must be added to $x^2 + 6x$ to make it into a perfect square.

What must be added to each of the following expressions to make it into a perfect square ?

1. $x^2 + 4x$ **5.** $x^2 - 3x$ **9.** $a^2 - \dfrac{3}{2}a$

2. $a^2 + 8a$ **6.** $x^2 + 20x$ **10.** $x^2 + x$

3. $x^2 - 12x$ **7.** $c^2 + 7c$ **11.** $x^2 + 2hx$

4. $p^2 - 14p$ **8.** $b^2 - \dfrac{1}{2}b$ **12.** $x^2 + \dfrac{b}{a}x$

Now consider $9x^2 + 12x$. If we want to make this into a perfect square we first take out the factor 9

i.e. $$9x^2 + 12x = 9\left(x^2 + \tfrac{12}{9}x\right)$$

$$= 9\left(x^2 + \tfrac{4}{3}x\right)$$

To complete the square within the bracket find the coefficient of x $\left(\text{i.e. } \tfrac{4}{3}\right)$, find half of it $\left(\tfrac{2}{3}\right)$, square this value $\left(\tfrac{4}{9}\right)$, and add it to the expression within the bracket.

Then $$9\left(x^2 + \tfrac{4}{3}x + \tfrac{4}{9}\right) = 9x^2 + 12x + 4$$

$$= (3x + 2)^2$$

which is in the form $(ax + b)^2$

EXERCISE 15c

What must be added to each of the following expressions to make it into a perfect square ?

1. $9x^2 + 12x$

2. $4x^2 + 12x$

3. $36a^2 + 60a$

4. $100x^2 - 60x$

5. $25x^2 - 20x$

6. $4x^2 + 20x$

7. $49a^2 - 28a$

8. $\dfrac{a^2}{4} - 2a$

9. $\dfrac{4}{9}a^2 - \dfrac{2a}{3}$

QUADRATIC EQUATIONS

Now consider the equation $(x + 1)^2 = 4$

If we take the square root of each side we get

$$x + 1 = \pm\sqrt{4}$$
$$= \pm 2$$

If $x + 1 = 2$
$$x = 1$$

and if $x + 1 = -2$
$$x = -3$$

These two values of x satisfy the equation $(x + 1)^2 = 4$.

EXERCISE 15d

Solve the equations:

1. $(x + 1)^2 = 9$

2. $(x - 2)^2 = 16$

3. $(x - 3)^2 = 25$

4. $(x + 6)^2 = 100$

5. $(x + 7)^2 = 1$

6. $(x - 1)^2 = 25$

7. $(x + 2)^2 = 49$

8. $(x - 5)^2 = 16$

9. $(x - 7)^2 = 4$

10. $(x + 4)^2 = 16$

11. $(x + 3)^2 = 25$

12. $(x - 9)^2 = 36$

13. $(x + 1)^2 = \frac{1}{4}$

14. $(x - 2)^2 = \frac{9}{4}$

15. $(x - \frac{1}{2})^2 = \frac{25}{4}$

Solve the equation $(2x + 3)^2 = 4$

Taking the square root of each side we get

$$2x + 3 = \pm 2$$

i.e. $\qquad 2x + 3 = 2 \qquad$ or $\qquad 2x + 3 = -2$

$\qquad \qquad 2x = -1 \qquad$ or $\qquad 2x = -5$

$\qquad \qquad x = -\frac{1}{2} \qquad$ or $\qquad x = -\frac{5}{2}$

Solve the equations:

16. $(2x - 1)^2 = 16$ **20.** $(7x + 2)^2 = 100$ **24.** $(4x - 3)^2 = 1$

17. $(3x + 2)^2 = 25$ **21.** $(2x + 1)^2 = 36$ **25.** $(9x - 5)^2 = 4$

18. $(5x - 1)^2 = 36$ **22.** $(3x - 4)^2 = 49$ **26.** $(5x + 3)^2 = 16$

19. $(3x - 4)^2 = 1$ **23.** $(5x + 2)^2 = 25$ **27.** $(7x - 5)^2 = 81$

SOLUTION OF QUADRATIC EQUATIONS
BY COMPLETING THE SQUARE

In Book 3A, Chapter 13, we were able to solve some quadratic equations by factorising the left-hand side and using the fact that if $A \times B = 0$ then either $A = 0$ or $B = 0$ (or A and B are both zero).

There are other equations which cannot be solved in this way because the left-hand side will not factorise. We can solve these equations by expressing them in the form $(x + a)^2 = c$ and taking the square roots of both sides. This is called the method of *completing the square*.

Consider the equation $\qquad x^2 - 6x + 2 = 0$

Subtract 2 from each side $\qquad x^2 - 6x = -2$

Complete the square on the LHS by adding 9 and add the same quantity to the RHS.

$$x^2 - 6x + 9 = -2 + 9$$

i.e. $\qquad \qquad (x - 3)^2 = 7$

Take the square root of each side $x - 3 = \pm 2.646$

$$x = 3 \pm 2.646$$

$$= 5.646 \text{ or } 0.354$$

i.e. $x = 5.65$ or 0.35 correct to 2 d.p.

These values of x are called the *roots* of the equation.

EXERCISE 15e

Solve the equation $x^2 + 5x - 3 = 0$ by completing the square.

$$x^2 + 5x - 3 = 0$$

$$x^2 + 5x = 3$$

$$x^2 + 5x + \frac{25}{4} = 3 + \frac{25}{4}$$

$$\left(x + \frac{5}{2} \right)^2 = \frac{12 + 25}{4}$$

$$\left(x + \frac{5}{2} \right)^2 = \frac{37}{4}$$

$$x + \frac{5}{2} = \pm \frac{\sqrt{37}}{2}$$

$$x = -\frac{5}{2} \pm \frac{6.083}{2}$$

$$= \frac{1.083}{2} \text{ or } \frac{-11.083}{2}$$

$$= 0.5415 \text{ or } -5.5415$$

i.e. $x = 0.54$ or -5.54 correct to 2 d.p.

Solve the following equations by completing the square.

1. $x^2 + 4x = 5$ **5.** $x^2 - 4x + 1 = 0$ **9.** $x^2 - 7x + 5 = 0$

2. $x^2 - 6x = 7$ **6.** $x^2 + 8x = 3$ **10.** $x^2 - x - 4 = 0$

3. $x^2 + 10x = 11$ **7.** $x^2 - 4x = 9$ **11.** $x^2 + 9x - 3 = 0$

4. $x^2 + 8x + 3 = 0$ **8.** $x^2 + 9x + 4 = 0$ **12.** $x^2 + 8x + 4 = 0$

Previously, when solving linear equations, we have frequently divided both sides by a non-zero number. When solving a quadratic equation by completing the square *always* divide both sides by the non-zero coefficient of x^2.

Solve equation $2x^2 + 6x - 5 = 0$ by completing the square.

$$2x^2 + 6x - 5 = 0$$

$$x^2 + 3x - \tfrac{5}{2} = 0$$

$$x^2 + 3x = \tfrac{5}{2}$$

$$x^2 + 3x + \tfrac{9}{4} = \tfrac{5}{2} + \tfrac{9}{4}$$

$$\left(x + \frac{3}{2}\right)^2 = \frac{10 + 9}{4}$$

$$\left(x + \frac{3}{2}\right)^2 = \frac{19}{4}$$

$$x + \frac{3}{2} = \pm\frac{\sqrt{19}}{2}$$

$$x = -\frac{3}{2} \pm \frac{4.359}{2}$$

$$x = \frac{1.359}{2} \quad \text{or} \quad \frac{-7.359}{2}$$

$$= 0.6795 \quad \text{or} \quad -3.6795$$

i.e. $x = 0.68$ or -3.68 correct to 2 d.p.

Solve the following equations by completing the square.

13. $2x^2 + 6x = 9$

14. $6x^2 - 12x = 5$

15. $4x^2 + 8x = 3$

16. $2x^2 - 3x - 4 = 0$

17. $3x^2 + 12x - 8 = 0$

18. $3x^2 - 5x = 1$

19. $5x^2 - 5x = 4$

20. $5x^2 + 8x + 2 = 0$

21. $4x^2 - 7x - 3 = 0$

22. $6x^2 - 5x - 1 = 0$

23. $7x^2 + 7x - 4 = 0$

24. $3x^2 - 9x = 2$

SOLUTION OF QUADRATIC EQUATIONS BY FORMULA

If we apply the method of completing the square to the general quadratic equation $ax^2 + bx + c = 0$, where a, b and c are positive or negative numbers, we can establish a formula for solving the equation.

Consider the general equation
$$ax^2 + bx + c = 0$$

Divide both sides by a
$$x^2 + \frac{b}{a}x + \frac{c}{a} = 0$$

Subtract $\frac{c}{a}$ from each side
$$x^2 + \frac{b}{a}x = -\frac{c}{a}$$

Complete the square on the LHS and add the same quantity to the RHS.
$$x^2 + \frac{b}{a}x + \frac{b^2}{4a^2} = -\frac{c}{a} + \frac{b^2}{4a^2}$$

Therefore
$$\left(x + \frac{b}{2a}\right)^2 = \frac{-4ac + b^2}{4a^2}$$

Take square roots of each side
$$x + \frac{b}{2a} = \pm\frac{\sqrt{b^2 - 4ac}}{2a}$$

Subtract $\frac{b}{2a}$ from each side
$$x = -\frac{b}{2a} \pm \frac{\sqrt{b^2 - 4ac}}{2a}$$

i.e.
$$x = \frac{-b \pm \sqrt{b^2 - 4ac}}{2a}$$

This is called the *formula* for solving quadratic equations. It gives values of x, or roots of the equation, for any given values of a, b and c (provided that $b^2 - 4ac$ is not negative).

Remember that a is the coefficient of x^2
b is the coefficient of x
c is the constant number term.

Since the two values of x are

$$-\frac{b}{2a} + \frac{\sqrt{b^2 - 4ac}}{2a} \quad \text{and} \quad -\frac{b}{2a} - \frac{\sqrt{b^2 - 4ac}}{2a}$$

the sum of the two roots is always $\left(\frac{-b}{2a}\right) + \left(\frac{-b}{2a}\right) = -\frac{b}{a}$

This provides a useful check that your answers are correct.

EXERCISE 15f

Use the formula to solve the equation $x^2 - 9x - 2 = 0$ giving your answers correct to two decimal places.

$$x^2 - 9x - 2 = 0$$

$$a = 1, \quad b = -9, \quad c = -2$$

$$x = \frac{-b \pm \sqrt{b^2 - 4ac}}{2a}$$

$$= \frac{-(-9) \pm \sqrt{(-9)^2 - 4(1)(-2)}}{2 \times 1}$$

$$= \frac{9 \pm \sqrt{81 + 8}}{2}$$

$$= \frac{9 \pm \sqrt{89}}{2}$$

$$= \frac{9 \pm 9.434}{2}$$

$$= \frac{18.434}{2} \quad \text{or} \quad \frac{-0.434}{2}$$

$$= 9.217 \text{ or } -0.217$$

$$\therefore \quad x = 9.22 \text{ or } -0.22 \quad \text{correct to 2 d.p.}$$

Check: Sum of roots is $9.22 + (-0.22) = 9$

and $\dfrac{-b}{a} = \dfrac{-(-9)}{1} = 9$

confirming the results.

Use the formula to solve the following quadratic equations.

1. $x^2 + 6x + 3 = 0$ **5.** $x^2 + 4x - 3 = 0$ **9.** $x^2 + 6x - 6 = 0$

2. $x^2 + 7x + 4 = 0$ **6.** $x^2 + 9x + 12 = 0$ **10.** $x^2 + 9x - 1 = 0$

3. $x^2 + 5x + 5 = 0$ **7.** $x^2 + 8x + 13 = 0$ **11.** $x^2 + 3x - 5 = 0$

4. $x^2 + 7x - 2 = 0$ **8.** $x^2 + 10x - 15 = 0$ **12.** $x^2 + 4x - 7 = 0$

13. $x^2 - 4x + 2 = 0$ **17.** $x^2 - 5x - 5 = 0$ **21.** $x^2 - 9x - 2 = 0$

14. $x^2 - 7x + 3 = 0$ **18.** $x^2 - 5x + 2 = 0$ **22.** $x^2 - 4x - 9 = 0$

15. $x^2 - 6x + 6 = 0$ **19.** $x^2 - 3x + 1 = 0$ **23.** $x^2 + 7x - 2 = 0$

16. $x^2 - 4x - 3 = 0$ **20.** $x^2 - 7x - 3 = 0$ **24.** $x^2 + 8x + 5 = 0$

Solve the equation $3x^2 + 7x - 2 = 0$ giving your answers correct to two decimal places.

$$3x^2 + 7x - 2 = 0$$

$$a = 3, \quad b = 7, \quad c = -2$$

$$x = \frac{-b \pm \sqrt{b^2 - 4ac}}{2a}$$

$$= \frac{-7 \pm \sqrt{7^2 - 4(3)(-2)}}{2 \times 3}$$

$$= \frac{-7 \pm \sqrt{49 + 24}}{6}$$

$$= \frac{-7 \pm \sqrt{73}}{6}$$

$$= \frac{-7 \pm 8.544}{6}$$

$$= \frac{1.544}{6} \text{ or } -\frac{15.544}{6}$$

$$= 0.257 \text{ or } -2.591$$

$$= 0.26 \text{ or } -2.59 \quad \text{correct to 2 d.p.}$$

Check: Sum of roots is $0.26 + (-2.59) = -2.33$

and $\dfrac{-b}{a} = \dfrac{-7}{3} = -2.33$ (to 2 d.p.)

confirming the results.

---

25. $2x^2 + 7x + 2 = 0$ **27.** $3x^2 + 7x + 3 = 0$

26. $2x^2 + 7x + 4 = 0$ **28.** $4x^2 + 7x + 1 = 0$

29. $5x^2 + 9x + 2 = 0$

30. $2x^2 - 7x + 4 = 0$

31. $4x^2 - 7x + 1 = 0$

32. $5x^2 - 9x + 2 = 0$

33. $3x^2 + 5x - 3 = 0$

34. $3x^2 + 9x - 1 = 0$

EXERCISE 15g

Solve the equation $4x^2 = 7x + 1$ giving your answers correct to two decimal places.

$$4x^2 = 7x + 1$$

(First arrange the equation in the form $ax^2 + bx + c = 0$)

$$4x^2 - 7x - 1 = 0$$

$$a = 4, \quad b = -7, \quad c = -1$$

$$x = \frac{-b \pm \sqrt{b^2 - 4ac}}{2a}$$

$$= \frac{-(-7) \pm \sqrt{(-7)^2 - 4(4)(-1)}}{2 \times 4}$$

$$= \frac{7 \pm \sqrt{49 + 16}}{8}$$

$$= \frac{7 \pm \sqrt{65}}{8}$$

$$= \frac{7 \pm 8.062}{8}$$

$$= \frac{15.062}{8} \quad \text{or} \quad \frac{-1.062}{8}$$

$$= 1.883 \text{ or } -0.133$$

$$= 1.88 \text{ or } -0.13 \quad \text{correct to 2 d.p.}$$

Check: Sum of roots is $1.88 + (-0.13) = 1.75$

and $\dfrac{-b}{a} = \dfrac{-(-7)}{4} = \dfrac{7}{4} = 1.75$

confirming the results.

1. $2x^2 = 8x + 11$

2. $4x^2 = 8x + 3$

3. $3x^2 = 3 - 5x$

4. $5x^2 = x + 3$

5. $4x^2 + 2 = 7x$

6. $3x^2 = 12x + 2$

7. $2x^2 = 3x + 1$

8. $4x^2 = 5 - 3x$

9. $3x^2 + 2 = 9x$

10. $6x^2 - 9x = 4$

11. $2x^2 = 5x + 5$

12. $3x^2 + 4x = 1$

13. $4x^2 = 4x + 1$

14. $3x^2 + 7x = 2$

15. $5x^2 = 5x - 1$

16. $8x^2 = x + 1$

EXERCISE 15h

In this exercise try the method of factorising first. If factors cannot be found use the formula.

1. $2x^2 + 3x - 2 = 0$

2. $3x^2 + 6x + 2 = 0$

3. $6x^2 + 7x + 2 = 0$

4. $2x^2 + 3x - 3 = 0$

5. $3x^2 - 8x + 2 = 0$

6. $3x^2 - 8x - 3 = 0$

7. $2x^2 - 3x - 3 = 0$

8. $8x^2 + 10x - 3 = 0$

9. $6x^2 + 7x - 2 = 0$

10. $4x^2 - 3x - 2 = 0$

11. $7x^2 + 8x - 2 = 0$

12. $5x^2 - 3x - 1 = 0$

13. $3x^2 = 7x - 2$

14. $11x^2 + 12x + 3 = 0$

15. $20x^2 = 3 - 11x$

16. $3x^2 - 14x + 15 = 0$

17. $5x^2 + 8x + 2 = 0$

18. $2x^2 = 7x + 3$

19. $2x^2 + 9x = 5$

20. $6x^2 = 5x + 2$

HARDER EQUATIONS

EXERCISE 15i

Solve the equation $\dfrac{1}{x+1} + \dfrac{2}{x-3} = 4$ giving your answers correct to two decimal places.

$$\frac{1}{x+1} + \frac{2}{x-3} = 4$$

Multiply both sides by $(x+1)(x-3)$

$$(x-3) + 2(x+1) = 4(x+1)(x-3)$$

$$x - 3 + 2x + 2 = 4(x^2 - 2x - 3)$$

$$3x - 1 = 4x^2 - 8x - 12$$

i.e.
$$4x^2 - 11x - 11 = 0$$

$$a = 4, \quad b = -11, \quad c = -11$$

$$x = \frac{-b \pm \sqrt{b^2 - 4ac}}{2a}$$

$$= \frac{11 \pm \sqrt{121 - 4(4)(-11)}}{8}$$

$$= \frac{11 \pm \sqrt{121 + 176}}{8}$$

$$= \frac{11 \pm \sqrt{297}}{8}$$

$$= \frac{11 \pm 17.234}{8}$$

$$= \frac{28.234}{8} \text{ or } \frac{-6.234}{8}$$

$$= 3.529 \text{ or } -0.779$$

i.e.
$$x = 3.53 \text{ or } -0.78 \quad \text{correct to 2 d.p.}$$

Solve the following equations; give answers correct to two decimal places.

1. $x + \dfrac{2}{x} = 11$

5. $\dfrac{2}{x+5} + \dfrac{3}{x-2} = 4$

2. $x - \dfrac{5}{x} = 3$

6. $\dfrac{2}{x} + \dfrac{1}{x+1} = 4$

3. $x + \dfrac{2}{x} = 7$

7. $\dfrac{3}{x-1} - \dfrac{2}{x+3} = 1$

4. $\dfrac{3}{x+2} - \dfrac{1}{x+4} = 2$

8. $\dfrac{5}{x} - 2x = 5$

PROBLEMS INVOLVING QUADRATIC EQUATIONS

EXERCISE 15j

The sum of two numbers is 13 and the sum of their squares is 97. Find the numbers.

Let one number be x then the other number is $13 - x$

and
$$(13 - x)^2 + x^2 = 97$$
$$169 - 26x + x^2 + x^2 = 97$$
$$2x^2 - 26x + 169 = 97$$
$$2x^2 - 26x + 72 = 0$$
$$x^2 - 13x + 36 = 0$$
$$(x - 4)(x - 9) = 0$$
$$x - 4 = 0 \text{ or } x - 9 = 0$$

i.e. $\qquad x = 4 \text{ or } x = 9$

If $\qquad x = 4, \ 13 - x = 9$

If $\qquad x = 9, \ 13 - x = 4$

The two numbers are therefore 4 and 9.

The following problems lead to quadratic equations that factorise.

1. The sum of two numbers is 13 and the sum of their squares is 85. Find them.

2. The difference between two positive numbers is 2 and the sum of their squares is 20. Find the numbers.

3. The sum of the squares of two consecutive positive numbers is 61. Find two numbers.

4. One side of a rectangle is 4 cm longer than the other. Find the sides if the area of the rectangle is 45 cm².

5. The perimeter of a rectangle is 26 cm and its area is 40 cm². Find the sides.

6. Two positive whole numbers differ by 3, and the difference between their squares is 39. If the smaller number is x form an equation in x and solve it to find the numbers.

7. The sides of a right-angled triangle are x cm, $(x+7)$ cm and $(x+8)$ cm. Find them.

8. A rectangle is 6 cm longer than it is wide. If its area is the same as that of a square of side 4 cm find its dimensions.

9. The sides of a right-angled triangle are x cm, $(x-2)$ cm and $(x-4)$ cm. Find them.

10. The hypotenuse of a right-angled triangle is 10 cm. Find the other two sides if their sum is 14 cm.

11. The product of two numbers is 84. If these numbers differ by 5, find them.

12. One number is 3 more than another. If their product is 88, find them.

13. The length of a rectangle is 5 cm more than its width. If the area of the rectangle is 36 cm² find its dimensions.

14. The base of a triangle is 5 cm more than its perpendicular height. If the area of the triangle is 42 cm² find
   a) the length of its base
   b) its perpendicular height.

15.

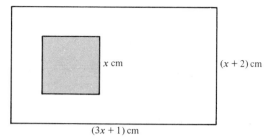

A square of side x cm is removed from a rectangular piece of cardboard measuring $(3x + 1)$ cm by $(x + 2)$ cm. If the area of card remaining is 62 cm² form an equation in x and solve it to find the dimensions of the original card.

16.

N is the midpoint of the base BC of a triangle ABC. If AB = AC, AN = x cm, BC = $(2x + 14)$ cm and AC = $(x + 8)$ cm form an equation in x and solve it. Hence find the length of the base and height of the triangle ABC.

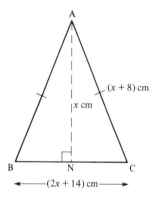

EXERCISE 15k

The following questions may lead to quadratic equations that do not factorise. Always check whether a quadratic equation will factorise before using the formula. If an answer is not exact give it correct to 3 s.f.

1. The sum of two numbers is 10 and the sum of their squares is 80. Find them.

2. The sum of two numbers is 9 and the difference between their squares is 60. Find them.

3. Find a number such that the sum of the number and its reciprocal is 20.

4. One side of a rectangle is 3 cm longer than another. Find the sides if the area of the rectangle is 20 cm².

5. Find the length of the hypotenuse of a right-angled triangle whose sides are x cm, $(x + 1)$ cm and $(x + 3)$ cm.

6. The parallel sides of a trapezium are $(x-2)$ cm and $(x+4)$ cm long. If the distance between the parallel sides is x cm and the area of the trapezium is 42 cm^2 find its dimensions.

7. A rectangular block is 2 cm wider than it is high and twice as long as it is wide. If its total surface area is 190 cm^2 find its dimensions.

8. Sally is x years old. Her mother's age is (x^2-4) years and her father is 6 years older than her mother. If the combined age of all three is 76 years form an equation in x and solve it. How old is her father?

HARDER PROBLEMS

EXERCISE 15I

A coach is due to reach its destination 30 kilometres away at a certain time. Its start is delayed by 18 minutes, but by increasing the average speed by 5 km/h the driver arrives on time. How long did the journey actually take? What was the intended average speed?

Let the intended average speed be x km/h. (The information can then be set out in table form taking care to work in compatible units.)

	Speed in km/h	Distance in km	Time in hours
Intended journey	x	30	$\dfrac{30}{x}$
Actual journey	$x+5$	30	$\dfrac{30}{x+5}$

Since the actual time is 18 minutes, i.e. $\dfrac{3}{10}$ hour, shorter than the intended time, then

$$\frac{30}{x} - \frac{30}{x+5} = \frac{3}{10}$$

Multiply both sides by $10x(x+5)$

$$300(x+5) - 300x = 3x(x+5)$$
$$100(x+5) - 100x = x^2 + 5x$$
$$100x + 500 - 100x = x^2 + 5x$$
$$0 = x^2 + 5x - 500$$

i.e.
$$x^2 + 5x - 500 = 0$$
$$(x+25)(x-20) = 0$$

∴
$$x = -25 \text{ or } 20$$

But -25 is unacceptable as the average speed has to be positive.

∴
$$x = 20$$

i.e. the intended speed is 20 kmph and the time actually

taken is $\dfrac{30}{20+5}$ hours $= \dfrac{30}{25}$ hours   i.e. 1 hour 12 min.

---

1. When its average speed increases by 10 m.p.h. the time taken for a car to make a journey of 105 miles is reduced by 15 minutes. Find the original average speed of the car.

2. Find the price of potatoes per kilogram if, when the price rises by 5 p per kg, I can buy 1 kg less for £2.10.

3. Tickets are available for a concert at two prices, the dearer ticket being £3 more than the cheaper one. Find the price of each ticket if a youth group can buy ten more of the cheaper tickets than the dearer tickets for £180.

4. In order to go on holiday Peter converts £300 into French francs. Had he bought three months earlier he would have received the same number of francs for £50 less, since the rate of exchange was two francs to the £ more then than on the day he bought them. Find the present rate of exchange.

5. From a piece of wire 42 cm long, a length $10x$ cm is cut off and bent into a rectangle whose length is one and a half times its width. The remainder is bent to form a square. If the combined area of the rectangle and square is 63 cm² find their dimensions.

6. The members of a club hire a coach for the day at a cost of £210. Seven members withdraw which means that each member who makes the trip must pay an extra £1. How many members originally agreed to go?

16 USING MONEY

MONEY

In many respects money is similar to any other commodity. Foreign currencies can be bought and sold in much the same way that cars can be bought and sold. Money can be lent and borrowed in a manner similar to hiring out or renting, say, a scaffold tower.

EXCHANGE RATES

When we shop abroad, prices quoted in the local currency often give us little idea of value so we tend to convert prices into sterling (£). To do this we need to know the exchange rate, i.e. how many units of the local currency are equivalent to one pound sterling.

For example, using an exchange rate of 11.50 Ff to £1

means that

$$£100 = 100 \times 11.50\,\text{Ff} = 1150\,\text{Ff}$$

and that

$$115\,\text{Ff} = £\frac{115}{11.5} = £10$$

A reasonable idea of cost is given by rounding off the exchange rate to make the arithmetic easy, but skill in mental arithmetic is useful! For example, a price of 250 Ff could be approximately converted to sterling by rounding $11.50\,\text{Ff} \equiv £1$ to $12\,\text{Ff} \equiv £1$.

Then

$$250\,\text{Ff} \approx £\frac{250}{12} = £21 \text{ to the nearest £1}$$

A more accurate conversion can be made using a conversion graph or a calculator.

EXERCISE 16a

If £1 is equivalent to 2630 Italian Lira (L), estimate the sterling equivalent of

a) 5000 L   b) 1000 L

    a)   (Approximating the exchange rate to 2500 L = £1 makes the arithmetic simple.)

$$5000\,L \approx £\frac{5000}{2500} = £2$$

    b)

$$1000\,L \approx £\frac{1000}{2500} = £\frac{2}{5} = 40\,p$$

This table gives the equivalent of £1 in various currencies.

£	French franc (Ff)	Spanish peseta (pta)	Italian lira (L)	Irish punt (pt)
1	11.60	220	2400	1.20

Use this table to   a) estimate   b) calculate (to the nearest penny) the sterling equivalent of

1. 200 Ff	**8.** 40 pta	**15.** 900 L
2. 20 Ff	**9.** 600 pta	**16.** 2.50 pt
3. 2500 Ff	**10.** 3810 pta	**17.** 4.80 pt
4. 5 Ff	**11.** 3000 L	**18.** 10.00 pt
5. 450 Ff	**12.** 10 000 L	**19.** 25.00 pt
6. 5000 pta	**13.** 250 L	**20.** 1.60 pt
7. 100 pta	**14.** 12 500 L	**21.** 3.70 pt

Use the table on page 289 to find how many pesetas are equivalent to 1 Ff.

From the table $11.60 \, \text{Ff} = 220 \, \text{pta}$

$$\therefore \qquad 1 \, \text{Ff} = \frac{220}{11.60} \, \text{pta}$$

$$= 18.97 \, \text{pta}$$

Use the table given at the beginning of the exercise to make the following conversions. Use your own judgement on how accurately your answers should be given.

22. 1 pt to Ff

23. 100 pta to Ff

24. 1 Ff to L

25. 100 L to pta

26. 10 pta to L

27. 3.50 pt to Ff

28. 650 pta to Ff

29. 25 Ff to L

30. 4500 L to pta

31. 280 pta to L

32. 84 Ff to L

33. 3800 L to pta

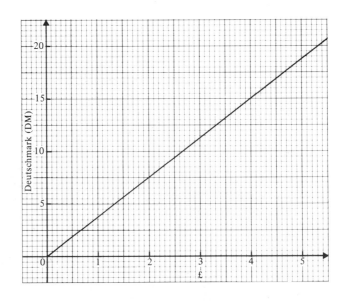

Use the conversion graph on p. 290 to find

34. £3 in DM **36.** £1.40 in DM **38.** 8 DM in £

35. £2.50 in DM **37.** 18 DM in £ **39.** 2.50 DM in £

40. Using an exchange rate of 1.40 U.S. dollars ($) to the pound sterling, make a conversion graph with a scale of 1 cm for £1 on the horizontal axis and a scale of 1 cm for $1 on the vertical axis. Use your graph to make the following conversions.

a) £2.50 to $ b) £4.75 to $ c) $4.20 to £.

EXCHANGE CROSS RATES

Institutions dealing with several foreign currencies need the exchange rate from any one foreign currency to any other. This information is published daily by the Financial Times in an exchange cross rate table.

EXCHANGE CROSS RATES

Oct. 11	£	$	DM	YEN	Ff	Sf	H Fl.	Lira	C$	B.Fr.
£	1.	1.412	3.753	303.0	11.45	3.078	4.233	2533.	1.929	76.10
$	0.708	1.	2.658	214.6	8.110	2.180	2.998	1794.	1.366	53.90
DM	0.266	0.376	1.	80.75	3.051	0.820	1.128	674.9	0.514	20.28
YEN (per 1000)	3.300	4.660	12.38	1000.	37.79	10.16	13.97	8358.	6.367	251.2
Ff (per 10)	0.873	1.233	3.277	264.6	10.	2.688	3.697	2212.	1.685	66.46
Sf	0.325	0.459	1.219	98.46	3.721	1.	1.375	822.9	0.627	24.73
H Fl.	0.236	0.334	0.887	71.59	2.705	0.727	1.	598.3	0.456	17.98
Lira (per 1000)	0.395	0.558	1.482	119.6	4.521	1.215	1.671	1000.	0.762	30.05
C$	0.518	0.732	1.945	157.1	5.935	1.595	2.194	1313.	1.	39.45
B Fr. (per 100)	1.314	1.855	4.931	398.2	15.05	4.044	5.562	3328.	2.535	100.

The left-hand column refers to 1 unit of currency unless otherwise stated.

Hence the first row of figures gives the equivalent of £1 in other currencies. The fourth row gives the equivalent of 1000 yen in other currencies, e.g. 1000 yen = 10.16 Sf (Swiss francs).

EXERCISE 16b

Use the table to give the following exchange rates:

1. £ to Canadian dollars (C $)
2. DM to Lira
3. Belgian Francs (Bf) to Yen
4. Dutch Florins (H Fl) to $
5. $ to Yen

6. Swiss Francs (Sf) to £
7. $ to £
8. Bf to $
9. DM to $
10. H Fl to £

BUYING AND SELLING FOREIGN CURRENCY

In the previous section we assumed that there is just one exchange rate between two currencies. For most holiday and business travel purposes it is reasonable to work with a single exchange rate. However, when changing sterling into foreign currency before going on holiday and then changing what is left of that currency back into sterling on return, we find that we have to deal with two exchange rates. High Street banks offering exchange display their rates under two headings: 'Bank Buys' and 'Bank Sells'. (Exchange rates vary slightly from bank to bank in much the same way that the cost of a packet of tea varies from shop to shop.) A typical display of exchange rates looks like this.

	Bank Buys	Bank Sells
Belgian f	78.60	74.80
French f	12.00	11.40
Deutchmarks	3.90	3.70
Italian Lira	2650	2500
Spanish pta	240	225
Swiss f	3.20	3.00
US $	1.40	1.32

This means that if we are exchanging French francs and sterling,

the bank will sell us French francs at 11.40 Ff to £1,

the bank will buy French francs from us at 12.00 Ff to £1.

Hence if we want to change £100 into French francs, then we will get

$$100 \times 11.40 \, Ff = 1140 \, Ff$$

If we want to change 1140 Ff into sterling then we will get

$$£\frac{1140}{12} = £95$$

In addition to the differential exchange rate, banks normally charge a commission on each transaction and this is typically 1% of the sterling value. Hence the first transaction of changing £100 into Ff is subject to a charge of 1% of £100, i.e. £1, so the cost of the 1140 Ff is £101.

Similarly the second transaction is subject to a charge of 1% of £95, i.e. 95 p. Therefore, for our 1140 Ff we would get £95 − 95 p, i.e. £94.05.

EXERCISE 16c

Use the table on page 292 to answer the following questions, giving your answers correct to *four* significant figures.

1. If I change £100 into deutchmarks, how many will I get?

2. If I come back from holiday with 250 DM how much sterling will the bank exchange them for?

3. What will it cost me in sterling to buy 1 000 000 lira from the bank?

4. How much will the bank pay me in sterling for U.S. $500?

5. How many pesetas will £500 buy?

6. A company exporting goods to Belgium is paid in Belgian francs and receives a cheque for 5000 Bf. How much in sterling will the company receive from its bank if the bank charges a commission of $1\frac{1}{2}$%?

7. Holiday flats in France are offered for rent at 2000 Ff a week. I rent one for two weeks and pay for it by writing out a cheque in French francs. How much does it cost me in sterling if the bank charges 1% commission?

8. Lesley Smith changed £100 into US $ for a business trip, but didn't spend any of the dollars. On return she changed the dollars back into sterling. If 1% commission was deducted on each transaction, how much did she lose?

9. A leather bag in an Italian shop is offered for sale at 125 000 L. A tourist, whose lira were bought in England at a charge of $1\frac{1}{2}$% commission, buys the bag. What is the cost in pounds of the bag to the tourist?

10. Mr and Mrs Edwards rented a flat in Spain for one week and paid 45 000 pta. On return they were given a refund of 5000 pta. If the bank charged them 1% commission on currency exchanges, find the cost in pounds of the rental, giving your answer correct to the nearest pound.

INVESTMENTS

Most people who have money that they do not need to spend immediately, invest it. Investing money means that the money is lent to an institution to use until such time as the owner wants it back. The institution pays the owner for the use of the money; this payment is called *interest.*

The most familiar forms of investment are Building Society accounts, Bank deposit accounts, Post Office savings accounts and Saving Certificates.

The interest payable on any investment is usually given as a percentage rate per annum, e.g. 8% p.a.

EXERCISE 16d

Give answers correct to the nearest penny.

An ordinary share account in a Building Society offers an interest rate of 8% p.a. payable half-yearly. If £100 is invested in this account and the interest is not withdrawn, find the amount in the account after 1 year.

Interest for the first 6 months is $\frac{1}{2}$ of 8% of £100 $= \dfrac{4}{100} \times £100$

$$= £4$$

(As the interest is not withdrawn, there is £104 invested for the second six months.)

Interest for the second 6 months is $\frac{1}{2}$ of 8% of £104 $= \dfrac{4}{100} \times £104$

$$= £4.16$$

After one year there is £108.16 in the account.

(Note that if the society paid the interest only yearly, the rate would have to be 8.16% p.a. to give the same amount. This rate of 8.16% is called the *compounded annual rate* (c.a.r.) and is often quoted by Building Societies along with their ordinary rate.)

1. £100 is invested in a Building Society. Assuming that no capital or interest is withdrawn find the amount in the account after 1 year when the rate of interest is

a) 5% p.a. payable half-yearly

b) 4% p.a. payable quarterly (every 3 months).

2. £100 is invested in a bank deposit account. If no capital or interest is withdrawn find the amount in the account after 1 year when the rate of interest is

a) 9% p.a. payable half-yearly

b) 9.2% p.a. payable yearly.

3. A savings account pays interest at the rate of 7% p.a. payable half-yearly. £200 is invested in this account. Find

a) the amount in the account after 1 year

b) the compounded annual rate.

4. £100 is invested in a Building Society at an interest rate of 10% p.a. payable half-yearly. Assuming that no capital or interest is withdrawn find

a) the amount in the account after 1 year

b) the compounded annual rate

c) the amount in the account after 2 years.

5. Savings Certificates are advertised by the statement '£100 becomes £120 in two years'. A building society offers interest at the rate of 9% p.a. payable half-yearly. If £100 was invested in the building society, what would this become after two years, and how does it compare with the Savings Certificates?

6. A Building Society offers a two-year bond of £1000 at a rate of 10% p.a. (A bond is a form of investment where the capital cannot be withdrawn during the term of the bond which in this case is two years.). The interest is paid direct to the investor every six months and cannot be credited to the bond. How much interest is paid over the two year term?

If, instead of buying one of these bonds, £1000 is invested in an account paying an interest rate of 8% p.a. payable half-yearly and if the interest is left in the account, how much would be in the account after two years?

7. A local authority offers a five-year bond of £1000, at a rate of 9% p.a. payable yearly. A bank offers a 'gold' savings account at a rate of 8.5% p.a. payable quarterly.

Tom Jones has £1000 to invest. If he wants to withdraw the interest only once a year, which of these two investments will give him the greater return?

INTEREST AND TAX

Interest from investments is called investment income and is subject to tax in much the same way as earned income is.

Building societies and banks pay interest which is net of tax to standard rate tax payers, i.e. they pay money direct to the Inland Revenue, so the investor does not have to pay standard rate tax on the interest. (However investors cannot claim this tax back from the Inland Revenue if they are non-tax payers.)

Sometimes interest rates are published in the form

8% p.a. net = 11.4% p.a. gross.

The gross rate is the interest the bank would have to pay to enable the investor to get the same amount of interest after paying the tax himself.

Some investments pay the gross rate of interest, so the investor has to pay any tax due.

EXERCISE 16e

A bank deposit account pays 8% p.a. net. Find the gross rate if standard rate tax is 35%.

On £100, 8% p.a. net gives interest of £8. The gross rate must give an amount which leaves £8 *after* tax of 35% is paid.

$$\therefore \qquad £8 = (100 - 35)\% \text{ of gross interest}$$

i.e. $$£8 = \frac{65}{100} \times (\text{gross interest})$$

$$£\frac{8 \times 100}{65} = \text{gross interest}$$

$\therefore$ gross interest on £100 is £12.31

$\therefore$ the gross interest rate is 12.31%.

1. A bank deposit account pays 7% p.a. net.
Find the gross rate if standard rate tax is 29%.

2. A building society account pays $9\frac{1}{4}$% p.a. net.
Find the gross rate if standard rate tax is 25%.

Complete the following table, giving answers correct to 2 d.p.

	Net rate of interest	Standard rate of income tax	Gross rate of interest
3.	5%	20%	
4.	6%	30%	
5.	8%		10%
6.	4%		5.5%
7.		40%	12.5%

8. A savings account offers interest of 8% p.a. gross, payable yearly. A building society offers interest of 6% p.a. net, payable yearly. If the standard rate of tax is 30%, which investment gives the bigger return to a tax payer and by how much?

9. An investor has £2000 to invest. He can choose from either an account paying 10% p.a. gross or an account paying 8% p.a. net. The standard rate of tax is 33%. Which choice gives him the better return and by how much if

a) he is a tax payer   b) he does not have to pay tax?

COMPOUND INTEREST

Suppose that £1000 is invested in an account that pays interest at 9% p.a. each year. If the capital and the interest are left in the account then, using the methods in the investment section successively for each year, we could find the amount in the account after, say, 7 years.

Alternatively, as the interest rate is 9% p.a. we can say that each year the account increases by $\frac{9}{100}$, or 0.09, of its value at the beginning of the year,
i.e.   its new value is   its original value + 0.09 of its original value,
i.e.   its new value is   1.09 of its original value.

Therefore the amount in the account after
    1 year is $1.09 \times £1000$
    2 years is $1.09 \times (1.09 \times £1000) = (1.09)^2 \times £1000$
    3 years is $1.09 \times (1.09)^2 \times £1000 = (1.09)^3 \times £1000$
    ......
    7 years is $(1.09)^7 \times £1000$

1.09^7 can be found using a calculator as follows

| 1 | . | 0 | 9 | y^x | 7 | = |

Therefore, after 7 years the amount is $1.828 \times £1000$

$$= £1828$$

The increase, £828, is called the *compound interest.*

1.09 is called the multiplying factor for 1 year and $(1.09)^7$ is the multiplying factor for 7 years.

If £500 is invested at an interest rate of 12% p.a. the multiplying factor for 1 year is 1.12.

After 9 years, the multiplying factor is $(1.12)^9$ i.e. in 9 years, the original £500 becomes

$$(1.12)^9 \times £500$$

$$= 2.773 \times £500$$

$$= £1387 \text{ (correct to the nearest £)}$$

APPRECIATION AND DEPRECIATION

Certain possessions such as houses and antiques tend to increase in value, or appreciate, as time passes. When the appreciation is expressed as a percentage rate per annum, the calculation is basically the same as for compound interest.

For example, suppose that a house bought for £20000 in 1980 appreciates each year by 5% of its value at the beginning of that year, then the yearly multiplying factor is 1.05,
i.e. in 1981 the house is worth $1.05 \times £20000$.

By 1990 the house has been appreciating for 10 years so its value then will be $(1.05)^{10} \times £20000$.

Other possessions such as cars and motor cycles tend to decrease in value, or depreciate, as time passes.

For example, if a car is bought for £8000 in 1980 and depreciates each year by 8% its value at the beginning of that year, then after 1 year it is worth 92% of £8000.

i.e. $0.92 \times £8000$

This time the multiplying factor is 0.92 for 1 year. After 7 years, the value of the car is $(0.92)^7 \times £8000$.

EXERCISE 16f

Give answers correct to the nearest penny when necessary.

1. £1000 is invested for 2 years at 5% p.a. compound interest. Find the compound interest.

2. £500 is invested for 2 years at 6% p.a. compound interest. How much is it then worth?

3. £3000 is invested for 3 years at 8% p.a. compound interest. Find the compound interest.

4. £1500 is invested for 3 years at 10% p.a. compound interest. Find the compound interest.

Use your calculator to find the following, giving your answers correct to four significant figures.

5. 1.12^8 **7.** 0.85^8 **9.** 1.11^5 **11.** 0.92^{10}

6. 1.08^8 **8.** 0.95^9 **10.** 1.15^{20} **12.** 0.87^{12}

13. £2000 is invested for 6 years at 8% p.a. compound interest. How much is it then worth?

14. £9000 is invested at 9% p.a. compound interest. How much is it worth after
a) 3 years   b) 10 years?

15. £8500 is invested at 5% p.a. compound interest. How much is it worth after 6 years?

16. Find the compound interest on £5000 invested for 8 years at   a) 9% p.a.
b) $8\frac{1}{2}$% p.a.

17. Find the compound interest on £6700 invested for 7 years at   a) 6.5% p.a.
b) 7.25% p.a.

18. Mr and Mrs Castrano buy a house for £40 000. What will it be worth, to the nearest £100, in   a) 2 years   b) 5 years, if it appreciates at 10% each year?

19. Miss Green buys a flat for £20 000. What will it be worth to the nearest £100, in   a) 4 years   b) 6 years, if its value increases each year by 8%?

20. John White pays £50 for a postage stamp for his collection. If its value appreciates by 15% each year what will it be worth in 5 years time?

21. A motorcycle bought for £1500 depreciates in value by 10% each year. Find its value, to the nearest £100, after   a) 3 years   b) 5 years.

22. When scientific calculators came on the market they were expensive, but in recent years have become much cheaper. In 1975 a good calculator cost £60. Assuming that the price of such a calculator reduced by 30% each year how much, to the nearest 10p, would one cost 4 years later?

23. The population of Mauritania is estimated to increase by 5% each year. If the population in 1980 was 1 500 000, estimate the population in 1990 giving your answer correct to the nearest hundred thousand.

COMPOUND GROWTH TABLE

In the previous exercise we needed to work out $(1.12)^8$ and other similar values. For convenience these multiplying factors are gathered together in Compound Growth Tables, and are frequently used commercially to show how sums of money, populations, sales, etc., grow at given percentage rates of increase over various periods of time.

The following table shows the multiplying factors for rates of growth from 6% to 15% over a 10 year period.

Rate of Growth	Number of Years									
	1	2	3	4	5	6	7	8	9	10
6%	1.060	1.124	1.191	1.262	1.338	1.419	1.504	1.594	1.689	1.791
7%	1.07	1.145	1.225	1.311	1.403	1.501	1.606	1.718	1.838	1.967
8%	1.08	1.166	1.260	1.360	1.469	1.587	1.714	1.851	1.999	2.159
9%	1.090	1.188	1.295	1.412	1.539	1.677	1.828	1.993	2.172	2.367
10%	1.100	1.210	1.331	1.464	1.611	1.772	1.949	2.144	2.358	2.594
11%	1.110	1.232	1.368	1.518	1.685	1.870	2.076	2.305	2.558	2.839
12%	1.120	1.254	1.405	1.574	1.762	1.974	2.211	2.476	2.770	3.106
13%	1.130	1.277	1.443	1.631	1.842	2.082	2.353	2.658	3.004	3.395
14%	1.140	1.300	1.482	1.689	1.925	2.195	2.502	2.853	3.252	3.707
15%	1.150	1.323	1.521	1.749	2.011	2.313	2.660	3.059	3.518	4.046

The table shows that the multiplying factor for a population growing at 8% for 8 years is 1.851 and that the multiplying factor for a sum of money growing at 12% for 9 years is 2.770.

EXERCISE 16g

In this exercise use the compound growth table on page 300. Give all answers correct to 4 significant figures.

What sum of money will £250 grow to if invested for 7 years at 12%?

From the table the multiplying factor is 2.211

i.e.   £250 will grow to 2.211 × £250

= £552.75

---

1. What sum will £10 grow to if invested for 3 years at 8%?

2. What sum will £20 grow to if invested for 6 years at 11%?

3. The population of Downtown is 2000. What will this increase to in 10 years if the growth rate is 6%?

4. Sally's wages have increased by 7% each year for the last 8 years. Eight years ago she was earning £100 per week. What is her present weekly wage?

5. Sales of Topmeat have increased steadily at 12% per year since it was introduced some years ago. Five years ago the company was selling 5000 tins a week. Find the present weekly sales.

Find the compound interest on

6. £300 invested for 10 years at 13%

7. £500 invested for 7 years at 7%

8. £700 invested for 9 years at 10%

9. £800 invested for 5 years at 12%

10. £400 invested for 8 years at 15%

Find the sum which, when invested for 8 years at 8% compound interest, will grow to £925.50.

£1 invested for 8 years at 8% compound interest will grow to £1.851

i.e.   £1.851 is what £1 will grow into in 8 years at 8%

∴   £925.50 is what £$\dfrac{925.50}{1.851}$, i.e. £500, will grow into in 8 years at 8%.

11.   Find the sum which, when invested for 6 years at 14%, will grow to £439.

12.   Find the sum which, when invested for 10 years at 9%, will grow to £1183.50.

13.   Find the sum which, when invested for 7 years at 6%, will grow to £1353.60.

14.   Find the population which, when growing at 12% for 5 years, will become 7048.

15.   My weekly wage has grown by a steady 6% each year for the past 10 years. I earn £89.55 each week now. What was my weekly wage 10 years ago?

In how many years will £800 grow to £1473.60 if invested at 13% compound interest?

If £800 grows to £1473.60 at 13% in the given time then

£1 grows to £$\dfrac{1473.60}{800}$ = £1.842 at 13% in the given time.

Using the table, go down to 13% and then go across until 1.842 is found.
This shows that the period of investment was 5 years.

In questions 16 to 20, how many years will it take

16. £300 to grow to £472.20 at 12% p.a. compound interest?

17. £500 to grow to £925.50 at 8% p.a. compound interest?

18. £800 to grow to £2046.40 at 11% p.a. compound interest?

19. a population of 75 000 to grow to 89 325 at an annual growth rate of 6%?

20. annual sales figures of 25 000 to grow to 62 550 at a steady annual growth rate of 14%?

21. If a population of 7400 grows to 14 245 in five years at a steady annual rate of x%, find x.

22. When £400 is invested for x years at 10% compound interest it amounts to £532.40. Find x.

CREDIT

Most of us need to borrow money at some time in our lives. When we buy goods or services and pay for them over an extended period of time we have obtained credit, (i.e. borrowed money). There are many ways of borrowing money and almost as many names for the different forms of loan.

MORTGAGES

Money borrowed to pay for a house or flat is called a mortgage. Mortgages are available from Building Societies, banks and some local authorities.

A mortgage is a long-term loan repayable, usually monthly, over several years. The deeds of the property are held by the lender for the duration of the mortgage. The charge for the loan is called interest. Mortgage interest rates are newsworthy because they fluctuate frequently and repayments form a large part of many people's expenditure.

When negotiating a mortgage, most people are interested in the size of the monthly repayments. These are frequently quoted per £1000 borrowed and vary with the interest rate and the number of years over which the loan is to be repaid.

HIRE PURCHASE, CREDIT SALES AND BANK LOANS

When buying items such as cars, furniture and larger electrical appliances, credit is available in the form of hire purchase, bank loans or credit sales.

Hire purchase differs from other forms of credit in that the article being purchased is hired from the hire purchase company and does not become legally the buyer's until the final payment is made.

A bank loan is a straightforward loan of money for a fixed term (typically between two and five years) with fixed monthly repayments. The goods that you buy with it are yours from the start, although security for the loan, such as deeds, or share certificates, is often required.

A credit sale is similar to a bank loan in that the goods are yours from the time of purchase and payment is usually by monthly instalments for a fixed term, typically 3 months to 2 years. Credit sale agreements are usually operated by the company supplying the goods, and security is not required. There is usually a charge for all these credit arrangements and by law this has to be clearly printed as an annual percentage rate (APR).

CREDIT CARDS

Credit cards like Access and Visa operate in a different way from traditional forms of credit. Each card holder is allocated a credit limit. Suppose that it is £500. This means that the card holder can use the card to pay for goods and services up to the total value of £500.

A statement is issued each month detailing how much has been spent and demanding a minimum payment towards this debt (this is usually £5 or 5% of the debt, whichever is the greater). The cardholder must pay at least the minimum figure but may pay more, or even pay off the full debt in which case no interest is charged.

If a part payment only is made there is a charge of about 2% per month on the outstanding debt from the date of the statement. This interest charge sounds low, but 2% per month is equivalent to 26.8% p.a.

EXERCISE 16h

(Where necessary work to the nearest penny.)

1. The Newtown Building Society offers a twenty year mortgage for monthly payments of £11.50 per £1000 borrowed. What are the monthly repayments on a £25 000 mortgage?

2. Jean and Michael Black want a mortgage of £20 000. The Redbrick Building Society offers them a 25-year mortgage with monthly repayments of £10.25 per £1000 borrowed. The Red Lion Bank offers them a 15-year mortgage with monthly repayments of £11.00 per £1000 borrowed.
For the full term of the mortgage, how much would they have to pay to
a) the building society   b) the bank?

3. Elizabeth Wood obtains a 95% mortgage on a house costing £30 000. Her monthly repayments are £13.60 per £1000 borrowed.

a) What are her monthly repayments?

b) If the mortgage runs for 25 years, what are her total repayments?

4. Mr and Mrs Smith have a mortgage on which the outstanding balance is £28 000. The interest rate is $12\frac{1}{2}$% p.a. What monthly payment would just cover the interest charges?

5. Zia Koren has a mortgage. When the outstanding debt is £9000 the interest rate rises from 10% p.a. to 11% p.a. The monthly repayments were £80 per month. Will it be necessary to increase these payments to stop the debt increasing, and if so by how much?

A motorcycle is priced at £2150. The hire purchase terms are 25% deposit plus 24 monthly payments of £82.50. Find the cost of the motorcycle if it is bought by hire purchase.

$$\text{Deposit} = 25\% \text{ of } £2150$$

$$= £537.50$$

$$\text{Total repayments} = 24 \times £82.50$$

$$= £1980$$

$$\text{Therefore HP cost} = £1980 + £537.50$$

$$= £2517.50$$

6. A hi-fi system is advertised at £520. If bought on HP, the terms are £50 deposit plus 18 monthly payments of £32. Find the HP cost.

7. A dining table and set of chairs is offered for £850 cash or on credit terms of 12 monthly payments of £81.50. Find the difference between the cash price and the credit price.

8. A freezer, advertised at £250, is offered for sale at either 2% discount for cash or on credit terms of 6 monthly repayments of £50.

a) How much is saved by paying cash?

b) What is the credit sale cost?

9. Mr Johnson wants to buy a car costing £7500, but cannot afford to pay cash. The following options for paying are available

a) A bank loan of £7500 repayable over 36 months at £281.38 per month.

b) A hire purchase agreement of a deposit of 25% of the cash price followed by 30 monthly repayments of £254.17

What is the cost of the car under each of these options?

10. A microcomputer package for business use is available under the following terms:

Either   a rental scheme costing £120 per month under which all service and repairs are free.

or   extended credit terms requiring payments of £220 per month for 18 months but any necessary service and repairs have to be paid for.

A company had one of these packages for three years, during which time it required service and repairs to the value of £400. The company chose the rental scheme. If instead they had chosen to buy the package on extended credit, would they have saved money or lost money and what is the difference?

11. When Tariq received his credit card statement the balance was £362.20. The minimum payment demanded is £5 or 5% of the balance whichever is the greater. What must Tariq pay?

12. James had a credit limit of £600 on his credit card. At the beginning of the month his balance was nil. He used the card during the month to pay £29.00 for petrol, £160 for clothes, £46 for a meal, £320 for a video recorder and £18.20 for records. He then offered his card for repayment of a garage bill of £150. The garage checked with the credit card company. Did they give the garage authorisation to accept the card? Give a reason for your answer.

17 GRADIENTS AND AREAS

TANGENT TO A CURVE

The graph shows the curve whose equation is $y = \frac{1}{2}x(5-x)$

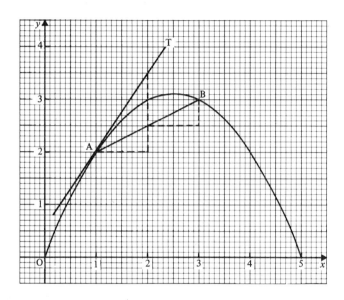

A line joining two points on a curve is a *chord*.

The line AB is a chord of the curve shown.

A line which touches a curve at a point is a *tangent* to the curve at that point.

In the diagram, AT is a tangent to the curve at A.

GRADIENTS

The gradient of a straight line is found by taking two points on the line and, moving left to right from one point to the other, evaluating the fraction

$$\frac{\text{distance moved up}}{\text{distance moved across}}$$

If you have to move down, the gradient is negative.

307

When choosing two points on the line, it is sensible to make the distance across a whole number of units.

If the coordinates of the two points are known, the gradient is given by

$$\frac{\text{difference in } y \text{ coordinates}}{\text{difference in } x \text{ coordinates}}$$

In the diagram, the gradient of the chord AB is $\frac{1}{2}$.

When moving along a curve, the gradient changes continuously. Imagine moving along the curve in the diagram, starting from O. When you get to A imagine that you stop following the curve and move on in a constant direction: you will move along the tangent AT.

Therefore

> the gradient of a curve at a point is defined as the gradient of the tangent to the curve at this point.

In the diagram, the gradient of the tangent AT is $\frac{3}{2}$, therefore the gradient of the curve at A is $\frac{3}{2}$.

Finding a gradient by drawing and measurement means that the curve must be accurately drawn and then the tangent must be positioned carefully. Using a transparent ruler helps and, as a rough guide, the tangent should be approximately at the same 'angle' to the curve on each side of the point of contact.

EXERCISE 17a

1. Draw x and y axes from -6 to $+6$ using a scale of 1 cm for 1 unit on both axes.

Plot the points A$(1, 1)$, B$(3, 2)$, C$(-4, -1)$, D$(-5, 1)$, E$(1, 6)$, F$(4, -5)$.

Find, where possible, the gradient of

a) AB   b) BC   c) DE
d) EF   e) AD   f) AE

2.

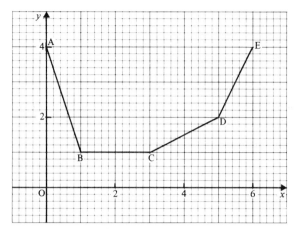

Find the gradient of the line

a) AB   b) BC   c) CD   d) DE

3. Copy and complete the table for $y = \dfrac{x^2}{10}$

x	0	1	2	3	4	5	6	7	8
y	0	0.1	0.4						

Use a scale of 2 cm for 1 unit on both axes and draw the curve.

a) P is the point on the curve where $x = 2$ and Q is the point on the curve where $x = 4$. Draw the chord PQ and find its gradient.

b) Draw, as accurately as possible, the tangents to the curve at the points where $x = 1, 4, 6$.

c) Find the gradients of the tangents to the curve at the points where $x = 1, 4, 6$.

4. Copy and complete the table for $y = \dfrac{10}{x}$, giving values of y correct to 1 d.p.

x	1	1.5	2	3	4	5	6	7	8
y	10	6.7	5	3.3	2.5				

Use a scale of 2 cm for 1 unit on both axes and draw the curve for values of x from 1 to 8.

a) A is the point on the curve where $x = 1$ and B is the point on the curve where $x = 4$. Find the gradient of the chord AB.

b) Find the gradients of the tangents to the curve at A and B.

c) Find the gradient of the tangent to the curve at the point where $x = 3$.

UNEQUAL SCALES

In the many practical applications of graphs it is rarely possible to have the same scales on both axes.

If the scales are not the same, care must be taken to read vertical measurements from the scale on the vertical axis and horizontal measurements from the scale on the horizontal axis.

EXERCISE 17b

The table shows the girth (w cm) of a pumpkin, n days after being fed with fertilizer.

n	1	2	3	4	5	6	7
w	15	17	20	24	29	35	41

Use a scale of 2 cm for 1 day and 2 cm for 10 cm of girth to draw the graph illustrating this information.

a) Find the gradient of the chord joining the points where $n = 3$ and $n = 7$ and interpret the result.

b) Find the gradient of the tangent at the point where $n = 3$ and interpret the result.

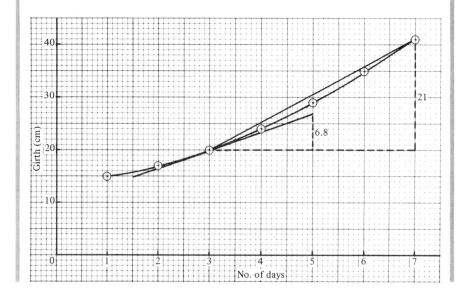

a) Gradient of chord $= \dfrac{21}{4} = 5.25$

This shows that the girth of the pumpkin is increasing by an average of 5.25 cm per day over the four-day period

b) Gradient of tangent $\approx \dfrac{6.8}{2} = 3.4$

This shows that the girth of the pumkin is increasing by 3.4 cm per day at the time of measurement on the third day.

---

1. The number of ripe strawberries on a particular strawberry plant were counted on Monday, Wednesday, Friday, Saturday and Sunday during one week. The results were recorded and plotted to give the following graph.

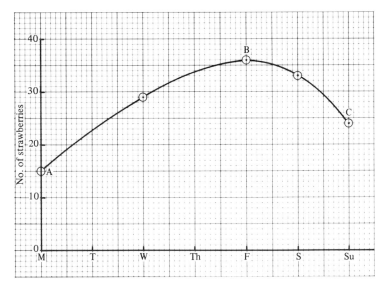

a) How many ripe strawberries were there on
   i) Wednesday   ii) Saturday?

b) How many strawberries were probably ripe on Thursday?

c) Find the gradient of the chord joining the points A and B and interpret the result.

d) Find the gradient of the chord joining B and C and interpret the result.

2. Draw a y-axis from 0 to 50 using a scale of 2 cm to 10 units and an x-axis from 0 to 6 using a scale of 2 cm to 1 unit.

Plot the points A(2, 10), B(1, 12), C(4, 25), D(0, 30), E(6, 45).

Find the gradients of the lines AB, BC, CD, DE.

3. The table shows the population of an island at 10 year intervals from 1900 to 1980.

Date (D)	1900	1910	1920	1930	1940	1950	1960	1970	1980
No. of people (N)	500	375	280	210	160	120	90	65	50

Using scale of $2\,\text{cm} \equiv 10$ years and $2\,\text{cm} \equiv 50$ people draw the graph illustrating this information.

a) Find the gradient of the chord joining the points on the curve where $D = 1910$ and $D = 1940$. Interpret your result.

b) Find the gradient of the tangent to the curve where $D = 1910$ and interpret the result.

4. The table shows the sales of 'Jampot' jam for 5 months following an advertising campaign.

Month (M)	1	2	3	4	5
Sales (No. of jars)	2000	2500	3500	5000	7000

Using a scale of $2\,\text{cm} \equiv 1$ month and $2\,\text{cm} \equiv 1000$ jars draw the graph illustrating this information.

a) Find the gradient of the tangent to the curve where $M = 2$ and interpret the result.

b) Find the gradient of the tangent to the curve where $M = 4$ and interpret the result.

5. Draw the graph of $y = x^3$ for values of x from 0 to 4 using a scale of $2\,\text{cm}$ for 1 unit on the x-axis and $1\,\text{cm}$ for 5 units on the y-axis.

Find the gradient of the curve at the points where

a) $x = 1$ b) $x = 3$

6. Copy and complete the following table of values for $y = 3x(4-x)$

x	0	1	2	2.5	3	4
y	0	9				

Using a scale of $2\,\text{cm}$ for 1 unit on the x-axis and $1\,\text{cm}$ for 1 unit on the y-axis draw the graph of $y = 3x(x-4)$.

Use the graph to find the gradient of the tangent to the curve where

a) $x = 2$ b) $x = 2.5$ c) $x = 3$

FINDING AREAS BOUNDED BY CURVES

It is sometimes necessary to find the area of a shape with curved boundaries. If the curve is part of a circle we can use the formula for the area of a circle. For other curves we can use approximate methods.

COUNTING SQUARES

For this method, the area required is drawn on squared paper using a suitable scale. (A grid of small squares produces a more accurate result than a grid of larger squares.)

For example, a survey of a village green resulted in the following scale drawing on 5 mm squared paper.

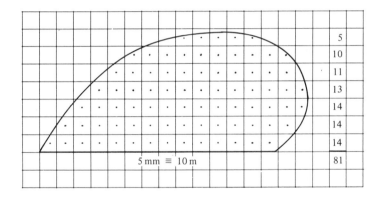

A square is included in the count if more than half of it is enclosed.

The number of squares enclosed in the diagram is 81.

Each square has a side of 5 mm and therefore an area of 25 mm².

Therefore the area of the scale drawing is 81×25 mm²

$$= 2025 \, \text{mm}^2$$

$$= 20.25 \, \text{cm}^2$$

Each square on the drawing represents a square of side 10 m on the village green (i.e. an area of 100 m²),

therefore the area of the green is 81×100 m²

$$= 8100 \, \text{m}^2$$

EXERCISE 17c

The following shapes are drawn on 5 mm squared paper. For each question find the area of

a) the shape drawn

b) the actual shape represented by the scale drawing (use the scale given).

1.

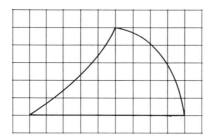

Scale: 5 mm ≡ 50 cm

2.

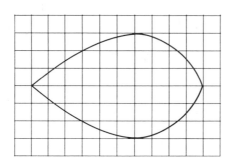

Scale: 5 mm ≡ 2 m

3.

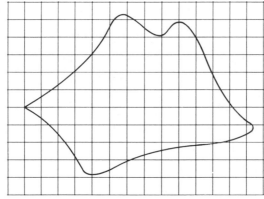

Scale: 5 mm ≡ 1 km

4. Draw a circle of radius 5 cm on 5 mm squared paper. Find the area of the circle

a) by counting squares

b) by using the formula $A = \pi r^2$ and a calculator.

AREAS UNDER GRAPHS

This graph shows the line $y = \frac{1}{3}(11 - 2x)$ drawn for values of x from 1 to 4.

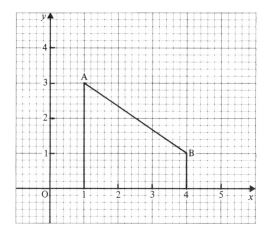

Consider the area enclosed by the section AB of the line, the x-axis, and the lines $x = 1$ and $x = 4$ (these are called *ordinates*). This is a trapezium and its area is given by

$$\frac{1}{2}(\text{sum of parallel sides}) \times (\text{distance between them})$$

The lengths of the parallel sides are 3 units and 1 unit, and the distance between them is 3 units.

Therefore the area under the line AB is

$$\frac{1}{2}(3 + 1) \times 3 \text{ square units} = 6 \text{ square units}.$$

Notice that, although 1 cm represents 1 unit on each axis we do not know what those units are, so the area is given as a number of square units.

When the scales on the two axes are different, care must be taken to read vertical lengths in units from the vertical scale and horizontal lengths in units from the horizontal scale.

EXERCISE 17d

Find the area between the line $y = 20 - 5x$, the x-axis and the ordinate $x = 1$.

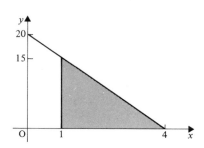

(Notice that a sketch of the graph is sufficient to show the area required; an accurate graph is not necessary.)

When $x = 1$, $y = 15$

The required area is a triangle of base 3 units and height 15 units.

Therefore area $= \frac{1}{2}(\text{base} \times \text{height})$

$= \frac{1}{2} \times 3 \times 15$ square units

$= 22.5$ square units.

In each case find, in square units, the area enclosed by the x-axis, the given line and the given ordinates.

1. $y = 1 + x$, $x = 0$, $x = 4$

2. $y = 5 - x$, $x = 0$

3. $y = 3 + 2x$, $x = 1$, $x = 5$

4. $y = 8 - 2x$, $x = 1$

5. $y = 20 - x$, $x = 4$ and $x = 10$

6. $y = 4 + 6x$, $x = 2$ and $x = 6$

7. $y = 15$, $x = 5$ and $x = 9$

8. $y = \frac{1}{2}(15 - x)$, $x = 3$ and $x = 12$

USING TRAPEZIUMS TO FIND THE AREA UNDER A CURVE

A curve can be approximated to by a series of straight lines.

To find the area between a curve and the *x*-axis, the area is divided into a convenient number of vertical strips. A chord is drawn across the top of each strip to give a set of trapeziums.

The sum of the areas of these trapeziums is then found and this is approximately equal to the required area.

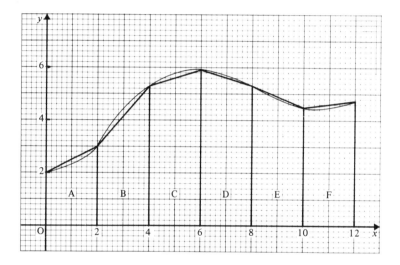

The area under this curve is divided into six strips each of width 2 units. Drawing the chords across the top of each strip gives six trapeziums.

Reading from the graph,

for trapezium A, the lengths of the parallel sides are 2 units and 3 units and the distance between them is 2 units,

therefore   area $A = \frac{1}{2}$ (sum of parallel sides) × (distance between them)

$$= \frac{1}{2}(2+3)\times(2) \text{ sq units}$$

$$= 5 \text{ sq units.}$$

Finding the sum of the areas of all six trapeziums gives the total area as

$$5 + 8.2 + 11.1 + 11.2 + 9.8 + 9.1 \text{ sq units}$$

$$= 54.4 \text{ sq units.}$$

Notice that all the strips are the same width. Although it is not necessary to use equal width strips, it is usually convenient.

EXERCISE 17e

Use the trapezium rule to find the area between each of the following curves and the *x*-axis. Use the given number of *equal width* strips.

1.

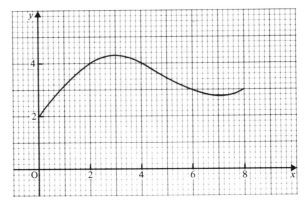

Use   a) two strips   b) four strips
Comment on your answers to parts (a) and (b).

2.

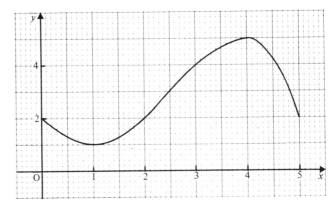

Use five strips.
Comment on the result if two strips were used.

3. A curve passes through the points given in the table

x	0	1	2	3	4
y	1	2	4	4	2

Use a scale of 1 cm for 1 unit on each axis and draw the curve for values of *x* from 0 to 4.
Use four strips to find approximately the area between this curve and the *x*-axis.

Use five strips to find, approximately, the area under the curve passing through the points given in the table for values of x from 1 to 11.

x	1	3	5	7	9	11
y	2	10	20	25	20	15

(A sketch graph is sufficient to give the necessary information.)

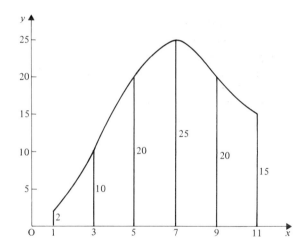

Area $\approx \frac{1}{2}(2+10)(2) + \frac{1}{2}(10+20)(2) + \frac{1}{2}(20+25)(2)$
$\qquad + \frac{1}{2}(25+20)(2) + \frac{1}{2}(20+15)(2)$ sq units

$= 12 + 30 + 45 + 45 + 35$ sq units

$= 167$ sq units

4. A curve goes through the points given in the table.

x	0	1	2	3	4
y	10	6	4	6	10

Draw a rough sketch of the curve.
Use four strips to find the area under the curve. Is your answer greater than or less than the true value?

5. Sketch the graph of the parabola $y = 3x^2$ from $x = 0$ to $x = 3$.

    a) Divide the area under the curve into three strips and mark the values of the ordinates on your sketch. Hence find, approximately, the area under your curve.

    b) Repeat part (a) using six strips.

6. A river is 40 metres wide and its depth was measured from one bank to the other bank at 5-metre intervals across its width. The values obtained are shown in the table.

Distance from bank (m)	0	5	10	15	20	25	30	35	40
Depth (m)	2	5	6	6.5	6	5	4	2.5	1.5

    a) Draw a rough sketch of the cross-section of the river.

    b) Use eight strips to estimate the area of the cross-section.

    c) The speed of the water at this point of the river is measured as 0.25 m/s. How many litres of water (to the nearest 10*l*) pass through this cross-section in one second?

    d) How many litres of water flow through the cross-section in one hour? Give your answer to the nearest 1000 litres.

18 TRAVEL GRAPHS

DISTANCE, SPEED AND TIME

In the metric system distance is measured in kilometres or metres while miles and feet are the most common Imperial units.

Time is measured in hours, minutes or seconds.

Speed is measured in distance units per time unit, e.g. a speed could be given in km per min.

Remember that

$$\text{distance} = \text{speed} \times \text{time}$$

and when using this formula, units must be consistent. For example, if a speed of 8 km/h and a time of 5 minutes are given then one quantity must be changed so that the time unit is the same for both.

EXERCISE 18a

Change a speed of 20 m/s to
a) m/min   b) km/h

a) $20 \, \text{m/s} = 20 \times 60 \, \text{m/min}$

$= 1200 \, \text{m/min}$

b) $20 \, \text{m/s} = \dfrac{20}{1000} \, \text{km/s}$

$= 0.02 \, \text{km/s}$

$= 0.02 \times 60 \times 60 \, \text{km/h}$

$= 72 \, \text{km/h}$

1. Change 33 km/h to   a) km/min   b) m/h

2. Change 100 m/s to   a) km/s   b) m/min

3. Change 40 m.p.h. to   a) miles/min   b) miles/s

<u>4.</u> Change 90 km/h to   a) km/s   b) m/s

<u>5.</u> Change 100 m/s to   a) m/h   b) km/h

Find, in metres per second, the average speed for a journey of 5 km in 4 minutes.

(For a speed in m/s, we need the distance in metres and the time in seconds.)

$$5 \text{ km} = 5000 \text{ m}$$
$$4 \text{ min} = 240 \text{ s}$$

Using $D = S \times T$ gives $S = \dfrac{D}{T}$

$$\text{Speed} = \frac{5000}{240} \text{ m/s}$$
$$= 20.8 \text{ m/s}$$

Find the unknown quantity in the following table, giving your answers in the units indicated.

	Distance	Speed	Time
6.	m	5 km/s	0.5 s
7.	90 m	m/s	4 min
8.	5 miles	m.p.h.	10 min
9.	2 km	4 km/h	min
10.	km	8 m/s	2 min

11.	miles	20 m.p.h.	5 min
12.	30 km	m/s	10 min
13.	120 m	m/s	5 min
14.	km	20 m/s	3 min
15.	1 km	25 m/s	s

DISTANCE-TIME GRAPHS

If a train travels from York at 40 km/h for 3 hours, stops for half an hour and then returns to York at 50 km/h, we can draw a graph to illustrate this journey.

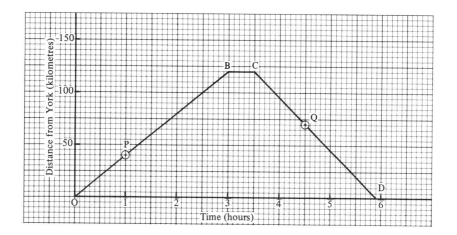

Time, as the independent variable, always goes on the horizontal axis.

OB represents the first part of the journey. The train starts at York and one hour later it is 40 km away. The point P, 1 hour along the time axis and 40 km along the distance axis is marked, and a straight line is drawn through O and P to B where the time is three hours.

BC represents the next part of the journey. The train is at rest for half an hour so a line, 0.5 h along the time axis, is drawn horizontally through B.

CD represents the last part of the journey. The train returns to York at 50 km/h so the line comes down, passing through the point Q which is 1 hour from C and 50 km down from C.

From the graph we see that the train returns to York 5.9 hours after leaving it.

EXERCISE 18b

Draw travel graphs to show the following journeys.

1. A cyclist leaves home and rides at 15 km/h for 2 hours. He rests for $\frac{1}{2}$ hour and then continues riding away from home at 12 km/h for $1\frac{1}{2}$ hours. He has another rest of $\frac{1}{2}$ hour before returning home, without stopping, at 20 km/h. For how long is the cyclist away from home ?
Use 1 cm for 1 hour and 1 cm for 10 km.

2. A bus leaves the bus station at 10.00 a.m. It travels at 30 km/h for $\frac{1}{2}$ hour, then stops for 12 minutes, before continuing at 40 km/h for 2 hours. How far has the bus travelled from the bus station now? Use 2 cm for 1 hour and 1 cm for 10 km.

3. A family travel by car from a town A to a town B 50 km away, in 48 minutes. They stop in town B for 1 hour and then continue to a village C, which is 70 km from B, at a speed of 50 km/h. The family stop at C for 2 hours and then return, without stopping, along the same route to A at 60 km/h. Use 1 cm for 1 hour and 1 cm for 20 km.

a) What is the speed of the journey from A to B ?

b) What is the time taken for the return journey from C to A ?

4. A bead is threaded on a straight wire, AB, 50 m long. The bead starts at A and travels to B at 20 m/s. It is held at B for 10 seconds and then returns to A at 10 m/s. Use 1 cm for 5 seconds and 1 cm for 10 metres.

What time elapses between the bead leaving A and returning to A ?

5. A funfair train runs on a track that is 200 m long. The track has three sections, A to B which is 50 m long, B to C which is 100 m long and C to the end D. The train travels at 5 m/s on the first section, 2 m/s on the second section and 4 m/s on the last section.
Use 1 cm for 10 seconds and 1 cm for 20 metres.

How long does the train take to travel the full length of the track ?

6. A lift, starting from the ground floor, travels up at 2 m/s for 12 seconds, to the top floor. It then stops for 20 seconds before descending to the ground floor. The descent takes 8 seconds.
Use 1 cm for 5 seconds and 1 cm for 5 metres.

a) How far above the ground floor is the top floor ?

b) What is the speed of the lift on its descent ?

AVERAGE SPEED

The average speed for a journey can be found from a distance-time graph.

Remember that

$$\text{average speed} = \frac{\text{total distance}}{\text{total time}}$$

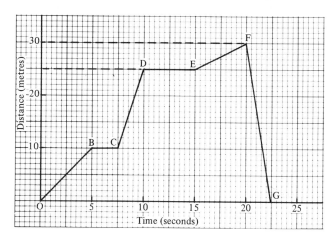

For the journey illustrated in the graph,

the section represented by O to D, a distance of 25 m, is covered in a time of 10 seconds,

therefore the average speed for this section is $\dfrac{25}{10}$ m/s = 2.5 m/s.

For the section represented by O to F, a distance of 30 m is covered in a time of 20 seconds,

therefore the average speed for this section is $\dfrac{30}{20}$ m/s = 1.5 m/s.

For the whole journey represented by O to G, a distance of 60 m (30 m away from the start and 30 m back again) is covered in 22.5 seconds.

therefore the average speed for the whole journey is $\dfrac{60}{22.5}$ m/s = 2.7 m/s.

EXERCISE 18c

Use the travel graphs drawn for Exercise 18b to find the average speed for
a) the journey during the first half of the time taken
b) the whole journey.

Give answers correct to 3 s.f. where necessary.

CURVED DISTANCE-TIME GRAPHS

When an object moves with constant speed, the distance-time graph representing its motion is a straight line. However, when an object moves so that its speed is constantly changing (e.g. a car or a big-dipper) then the distance-time graph representing its motion is a curved line. To draw such a graph, we need a relation between the time and the distance travelled.

Consider, for example, a rocket fired from the ground so that its distance (*d* metres) from the launching pad after *t* seconds is given by

$$d = 80t - 5t^2$$

Taking values of *t* from 0 to 8, at two-second intervals, gives the following table.

t	0	2	4	6	8
$80t$	0	160	320	480	640
$5t^2$	0	20	80	180	320
d	0	140	240	300	320

Plotting these values and drawing a smooth curve through the points gives the distance–time graph.

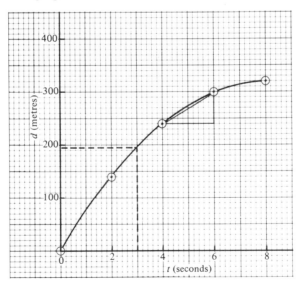

From this graph we can see that

3 seconds after launching, the rocket is 195 m above the launch pad.

We can also find the average speed of the rocket over any interval of time.

Consider, for example, the motion during the fifth and sixth seconds. The rocket moves $(300 - 240)\,\text{m}$, i.e. $60\,\text{m}$, in these 2 seconds. Therefore the average speed for the interval from $t = 4$ to $t = 6$

$$\text{is } \frac{60}{2}\,\text{m/s} = 30\,\text{m/s}$$

Notice that the chord joining the points on the curve where $t = 4$ and $t = 6$, has a gradient of $\dfrac{60}{2} = 30$.

EXERCISE 18d

1. A car moves away from a set of traffic lights. The table shows the distance, d metres, of the car from the lights after t seconds.

t	0	1	2	3	4	5
d	0	2	8	18	32	50

Draw the distance–time graph using scales of $1\,\text{cm} \equiv \frac{1}{2}$ second and $1\,\text{cm} \equiv 5\,\text{m}$.

From your graph find

a) the distance of the car from the lights after $2\frac{1}{2}$ seconds

b) the average speed of the car during the 2nd second

c) the average speed of the car during the first five seconds.

2. A rocket is launched and the table shows the distance travelled, d metres, after a time t seconds from lift-off.

t	0	1	2	3	4	5
d	0	5	40	135	320	625

Draw the distance–time graph using scales of $2\,\text{cm} \equiv 1$ second and $1\,\text{cm} \equiv 100\,\text{m}$.

Use your graph to find

a) the distance of the rocket from the launch pad $4\frac{1}{2}$ seconds after lift-off

b) the average speed of the rocket for the first 4 seconds of its journey

c) the average speed of the rocket during the fourth second.

VELOCITY

Consider a bead threaded on a straight wire AB.

If we are told that the bead is moving along the wire at 5 m/s, we know something about the motion of the bead but we do not know which way the bead is moving.

If we are told that the bead is moving from A to B at 5 m/s we know *both* the direction of motion *and* the speed of the bead.

Velocity is the name given to the quantity that includes the speed *and* the direction of motion.

When an object moves along a straight line, like the bead, there are only two possible directions of motion. In this case a positive sign is used to indicate motion in one direction and a negative sign is used to indicate motion in the opposite direction.

Taking the direction A to B as positive, we can illustrate velocities of +5 m/s and −5 m/s on a diagram.

Velocity = +5 m/s   Velocity = −5 m/s

EXERCISE 18e

1. For each of the following statements state whether it is the velocity or the speed of the object that is given.

a) A train travels between London and Watford at 70 km/h.

b) A train travels from London to Watford at 70 km/h.

c) A ball rolls down a hill at 5 m/s.

d) A ball rolls along a horizontal groove at 3 m/s.

e) A lift moves between floors at 2 m/s.

f) A lift moves up from the ground floor at 2 m/s.

g) A ferry crosses the channel from Dover to Calais at 15 knots (nautical miles per hour).

2. A bead moves along a horizontal wire AB. Taking the direction from A to B as positive, draw a diagram to illustrate the motion of the bead if its velocity is

a) $-2\,\text{m/s}$ b) $4\,\text{m/s}$ c) $-10\,\text{m/s}$ d) 0

3.

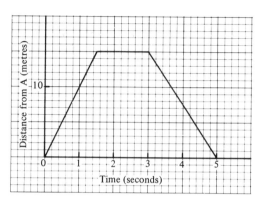

The graph illustrates the motion of a bead along a straight wire AB as the bead moves from A to B, stops at B, and moves back to A again.

Taking the direction $\overrightarrow{\text{AB}}$ as a positive, find

a) the speed of the bead as it moves from A to B

b) the velocity of the bead as it moves from A to B

c) the speed of the bead as it moves from B to A

d) the velocity of the bead as it moves from B to A

e) the average speed for the whole motion.

4.

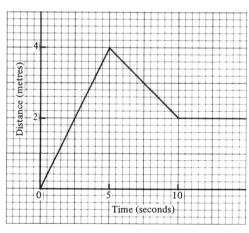

The graph illustrates the motion of a ball rolling in a straight line along horizontal ground. Taking the direction of the first part of the motion as positive, describe the motion of the ball, giving its velocity for each section of the motion.

5.

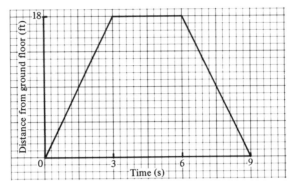

The graph above illustrates the motion of a lift which travels from the ground floor to the first floor and then returns to the ground floor. Taking the upward direction as positive, state which of the following statements *must* be true.

a) The velocity of the lift is the same on both the upward and downward journeys.

b) On the downward journey the speed is 6 ft/s.

c) The average speed of the lift between leaving the ground floor and returning to it, is zero.

d) On the upward journey the velocity of the lift is 6 ft/s.

e) The velocity of the lift is zero for three seconds.

FINDING VELOCITY FROM A DISTANCE-TIME GRAPH

This distance-time graph illustrates the motion of a ball thrown upwards from the ground.

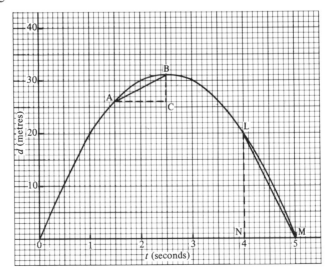

Taking the upward direction as positive we see that, up to the point where $t = 2\frac{1}{2}$, the distance of the ball from the ground is increasing, i.e. the ball has a positive velocity (the ball is going up).

From $t = 2\frac{1}{2}$ to $t = 5$, the distance of the ball from the ground is decreasing, i.e. the ball has a negative velocity (the ball is going down).

Average velocity is defined as $\dfrac{\text{increase in distance}}{\text{time taken}}$

Therefore the average velocity of the ball in the interval from $t = 1.5$ to $t = 2.5$ is

$$\frac{\text{increase in distance from } t = 1.5 \text{ to } t = 2.5}{2.5 - 1.5} \text{ m/s} = \frac{5}{1} \text{ m/s}.$$

On the graph this is represented by $\dfrac{\text{BC}}{\text{AC}}$ which is the gradient of the chord AB.

Similarly the average velocity of the ball from $t = 4$ to $t = 5$ is

$$\frac{\text{the increase in distance in this time}}{5 - 4} \text{ m/s} = \frac{-20}{1} \text{ m/s}$$

On the graph this is represented by $\dfrac{-\text{LN}}{\text{NM}}$ which is the gradient of the chord LM.

Therefore on a distance–time graph,

> the average velocity from t_1 to t_2 given by the gradient of the chord joining the points on the curve where $t = t_1$ and $t = t_2$.

The average velocity during any time interval can now be found using this fact, as the following example shows.

From $t = 1\frac{1}{2}$ to $t = 4$ the average velocity is given by the gradient of the chord AL, i.e.

$$\frac{-6}{2.5} \text{ m/s} = -2.4 \text{ m/s}$$

VELOCITY AT AN INSTANT

Consider again the distance–time graph on p. 330.

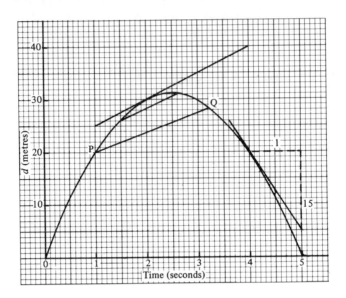

Suppose that we want to find the velocity of the ball at the instant when $t = 2$.

We can get an approximate value for this velocity by finding the average velocity from, say, $t = 1$ to $t = 3.2$, i.e. by finding the gradient of the chord PQ.

A better approximation is obtained by taking a smaller interval of time, say $t = 1.5$ to $t = 2.6$, i.e. by making the ends of the chord closer together.

The best answer is obtained when the interval of time is as small as possible, i.e. when the ends of the chord coincide. When this happens the chord becomes a tangent to the curve at the point where $t = 2$.

Therefore the velocity of the ball at the instant when $t = 2$ is given by the gradient of the tangent to the curve at the point where $t = 2$.

By drawing and measurement, the velocity is 5 m/s. Similarly, at the point where $t = 4$, the gradient of the tangent is found to be -15. Therefore the velocity when $t = 4$ is -15 m/s.

In general

> the velocity at the instant when $t = T$ is given by the gradient of the tangent to the distance–time curve at the point where $t = T$.

EXERCISE 18f

1. Use the graph drawn for question 1 in Exercise 18d p. 327 to find
   a) the average velocity during the first three seconds
   b) the velocity of the car 2 seconds after leaving the lights
   c) the velocity of the car 4 seconds after leaving the lights.

2. Use the graph drawn for question 2 of Exercise 18d to find
   a) the average velocity of the rocket during the time from $t = 2$ to $t = 4$
   b) the average velocity of the rocket over the interval $t = 2$ to $t = 3$
   c) the velocity of the rocket when $t = 2$
   d) the velocity of the rocket when $t = 2.5$.

3. The table shows the distance, d metres, of a ball from its starting position, t seconds after being thrown into the air.

t	0	1	2	3	4	5	6
d	0	25	40	45	40	25	0

Use scales of $2\,\text{cm} \equiv 1$ second and $1\,\text{cm} \equiv 5\,\text{m}$ and draw the graph of d against t.

From your graph find
   a) when the ball returns to the starting point
   b) the average velocity of the ball from $t = 1$ to $t = 2$
   c) the average velocity of the ball from $t = 1$ to $t = 1.5$
   d) the velocity of the ball when $t = 1$
   e) the velocity of the ball when $t = 4$
   f) the velocity of the ball when $t = 5$.

4. Use the graph at the foot of page 330 to find

    a) the velocity when $t = 1$

    b) the average velocity during the first second

    c) the average velocity during the first four seconds

    d) the greatest height of the ball

    e) the average velocity for the time between $t = 3$ and $t = 5$.

5. A particle moves in a straight line so that t seconds after leaving a fixed point O on the line, its distance, d metres, from O is given by

$$d = 8t - 2t^2$$

    a) Copy and complete the following table

t	0	1	2	3	4
$8t$	0	8			
$-2t^2$	0	-2			
d	0	6			

    b) Use scales of $2\,\text{cm} \equiv 1\,\text{second}$ and $1\,\text{cm} \equiv 2\,\text{m}$ and draw the distance–time graph.

    c) From your graph find the velocity of the particle when $t = 2$ and when $t = 3$.

    d) What is the greatest distance of the ball from the ground?

ACCELERATION

When the speed of a moving object is increasing we say that the object is accelerating.

If a train moves away from a station A and accelerates from rest so that its speed increases by $2\,\text{m/s}$ each second then

    1 second after leaving A the train has a speed of $2\,\text{m/s}$
    2 seconds after leaving A the train has a speed of $4\,\text{m/s}$
    3 seconds after leaving A the train has a speed of $6\,\text{m/s}$.

The train is said to have an acceleration of $2\,\text{m/s}$ per second and this is written as $2\,\text{m/s/s}$ or $2\,\text{m/s}^2$.

If the speed of the train decreases it is said to be decelerating.

Suppose that the speed of the train decreases by 1 m/s each second, then we say that the deceleration is 1 m/s per second or 1 m/s².

We can also say that the train has an acceleration of −1 m/s², i.e. a deceleration is a negative acceleration.

Consider a car that accelerates from rest at 5 m/s² for 10 seconds and then decelerates at 2 m/s² back to rest.

An acceleration of 5 m/s² means that the speed of the car increases by 5 m/s each second. Therefore after 10 seconds its speed is 50 m/s. A deceleration of 2 m/s² means that the speed of the car reduces by 2 m/s each second. Therefore, the speed of 50 m/s is reduced by 2 m/s each second, and this means that the car takes 25 seconds to stop.

EXERCISE 18g

1. A train accelerates from rest at 1 m/s² for 30 seconds. How fast is the train moving at the end of the 30 seconds?
If the train now decelerates back to rest at 0.5 m/s² how long does it take for the train to stop?

2. A train accelerates from rest at 2 m/s². How fast is the train moving after
a) 2 seconds   b) 30 seconds   c) 1 minute?

3. A bus moves away from rest at a bus stop with an acceleration of 4 m/s² for 5 seconds; it then has to decelerate to rest at 2 m/s². How long after leaving the bus stop is the bus again stationary?

4. The speed of a lift increases from 6 m/s to 20 m/s in 7 seconds. Find the acceleration.

5. A train accelerates from rest at 0.5 km/minute². How fast (in km/h) is the train moving after
a) 3 minutes   b) 10 minutes   c) 45 seconds?

6. The speed of a car increases from 10 km/h to 80 km/h in 5 seconds. Find the acceleration.

7. Find, in m/s², the acceleration of a motor bike when its speed increases from 10 km/h to 50 km/h in 4 seconds.

VELOCITY-TIME GRAPHS

A car accelerates from rest at $5 \, \text{m/s}^2$ for 6 seconds and then travels at a constant speed for 10 seconds after which it decelerates to rest at $3 \, \text{m/s}^2$.

This information can be shown on a graph by plotting velocity against time. This is called a velocity-time graph.

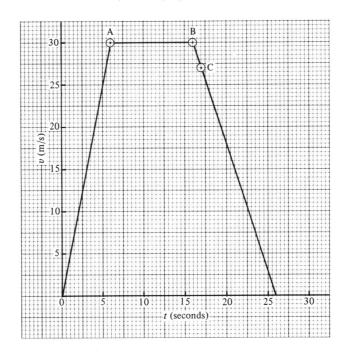

After 6 seconds, the car is moving at $30 \, \text{m/s}$, so we draw a straight line from O to the point A, where $t = 6$ and $v = 30$.
Notice that the gradient of OA represents the acceleration.
The line AB (zero gradient) represents the car moving at constant speed.

The last section of the journey is represented by the line drawn from B through C to the time axis, where C is 1 unit along the time axis and 3 units down the velocity axis from B.
Notice that the gradient of BC is -3 and this represents the deceleration of $3 \, \text{m/s}$.

> In general, acceleration is represented by the gradient of
> the velocity-time graph.

EXERCISE 18h

1.

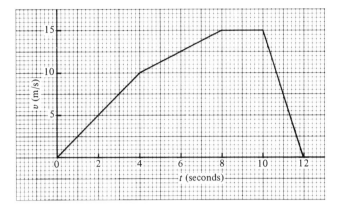

This velocity–time graph illustrates the journey of a car between a set of traffic-lights and a zebra crossing.

a) What is the car's acceleration for the first 4 seconds?

b) What happens when $t = 4$?

c) For how long is the car moving at a constant speed?

d) For how long is the car braking?

e) What is the deceleration of the car?

f) For how long is the car moving?

2.

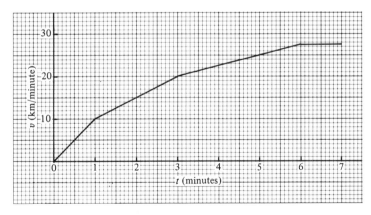

This velocity–time graph illustrates the first 7 minutes of the flight of a rocket.

a) What is the initial acceleration of the rocket?

b) What is the speed of the rocket 2 minutes after its launch?

c) What steady speed is obtained by the rocket?

d) What is the acceleration of the rocket during the fourth minute of its flight?

3.

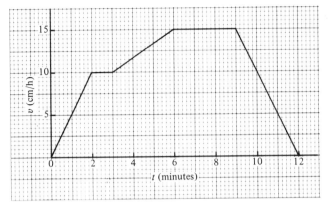

This velocity-time graph illustrates the journey of a train between two stations.

a) What is the acceleration for the first 2 minutes?

b) What is the greatest speed of the train?

c) For how long is the train travelling at constant speed?

d) What is the acceleration during the third minute?

e) What is the deceleration of the train?

Draw a velocity-time graph to illustrate the following journeys.
Use scales of $1\,\text{cm} \equiv 2$ seconds and $1\,\text{cm} = 5\,\text{m/s}$.

4.   A car accelerates steadily from rest reaching a speed of $12\,\text{m/s}$ in 10 seconds.

5.   A train accelerates from rest reaching a speed of $8\,\text{m/s}$ in 5 seconds and then immediately decelerates to rest in 4 seconds.

6.   A motorbike accelerates from rest to a speed of $20\,\text{m/s}$ in 4 seconds, maintains this steady speed for 8 seconds and then decelerates to rest in 5 seconds.

Sketch a velocity-time graph for each of the following journeys.

7.   A bullet is fired into a block of wood at $100\,\text{m/s}$ and comes to rest 3 seconds later.

8.   A car accelerates from rest at $2\,\text{m/s}^2$ for 5 seconds, then moves with constant speed for 15 seconds before decelerating back to rest at $4\,\text{m/s}^2$.

9.   A train accelerates from rest at $1\,\text{m/s}^2$ for 3 seconds, $2\,\text{m/s}^2$ for 3 seconds and then $5\,\text{m/s}^2$ for 5 seconds.

10.   A car accelerates from rest at $10\,\text{m/s}^2$ for 2 seconds, $5\,\text{m/s}^2$ for 5 seconds and then $2\,\text{m/s}^2$ for 3 seconds. The car then travels at constant speed for 10 seconds before decelerating at $8\,\text{m/s}^2$ back to rest.

11. A bullet is fired at 50 m/s into sand which retards the bullet at 10 m/s².

12. A car travels at 30 m/s for 5 seconds, then decelerates at 4 m/s² for 3 seconds and travels at constant speed for another 10 seconds.

13. A block of wood is thrown straight down into the sea. The wood enters the water at 50 m/s and sinks for 6 seconds.

FINDING THE DISTANCE FROM A VELOCITY-TIME GRAPH

This velocity-time graph shows a car accelerating from a speed of 4 m/s to a speed of 20 m/s in 9 seconds.

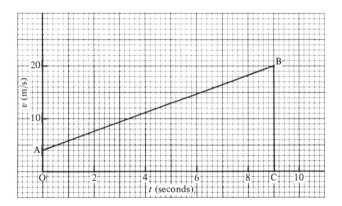

The time for which the car is accelerating is represented by the line OC. The speed of the car increases from 4 m/s to 20 m/s.

Therefore the average speed of the car is $\frac{1}{2}(4 + 20)$ m/s.

On the graph, this average speed is represented by $\frac{1}{2}(OA + BC)$.

Now   average speed $= \dfrac{\text{distance covered}}{\text{time}}$

so   distance covered $=$ average speed $\times$ time

On the graph this is represented by $\frac{1}{2}(OA + BC) \times OC$.

This is the area of the trapezium OABC, i.e.

> the distance travelled is represented by the area under the velocity-time graph.

EXERCISE 18i

The velocity-time graph illustrates a train journey between two stations.

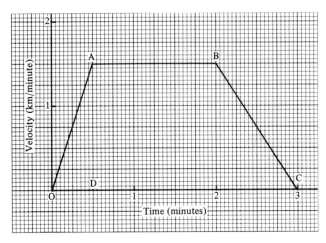

a) What is the maximum speed of the train in km/min?
b) What is the train's acceleration in the first half minute?
c) How far does the train travel in the first 30 seconds?
d) What is the distance between the stations?

a) From the graph, the maximum speed is 1.5 km/min

$$= 1.5 \times 60 \text{ km/h}$$

$$= 90 \text{ km/h}$$

b) The acceleration is given by the gradient of OA, which is $\dfrac{1.5}{0.5} = 3$

Therefore the acceleration is 3 km/minute^2.

c) The distance travelled in the first 30 seconds is given by the area of $\triangle$OAD

$$\text{area } \triangle\text{OAD} = \tfrac{1}{2}(\text{OD}) \times (\text{AD})$$

$$= 0.5 \times (0.5) \times (1.5)$$

$$= 0.375$$

Therefore the train travels 0.375 km in the first 30 seconds.

d) The distance between the stations is represented by the area of trapezium OABC.

$$\text{area OABC} = \tfrac{1}{2}(\text{OC} + \text{AB}) \times \text{AD}$$

$$= 0.5(3 + 1.5) \times 1.5$$

$$= 0.5 \times 4.5 \times 1.5$$

$$= 3.375$$

Therefore the distance between the stations is 3.375 km.

1. The velocity–time graph represents a car journey between two sets of traffic lights.

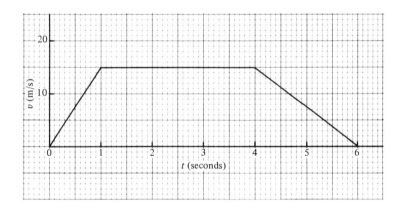

a) What is the acceleration of the car?

b) What is the deceleration of the car?

c) For how long does the car accelerate?

d) How many metres does the car travel while braking?

e) Find the distance between the two sets of lights.

2. Use the graphs for questions 1, 2 and 3 of Exercise 18h to find

a) the distance covered by the car in question 1

b) the distance travelled by the rocket in the first 3 minutes in question 2

c) the distance travelled by the train in the first 2 minutes in question 3. (Be careful with the units.)

3. The velocity–time graph represents a bullet fired into a 'wall' made up of a layer of sand followed by a layer of wood and then a layer of brick.

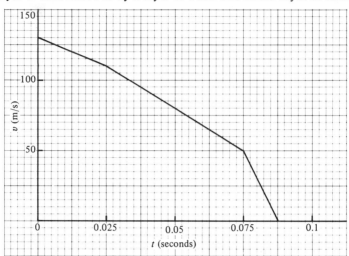

a) Find the deceleration of the bullet as it passes through the layer of sand.

b) Find the depth of the sand.

c) Find the deceleration of the bullet as it passes through the layer of wood

d) Find the depth of the layer of wood.

e) What retardation of the bullet does the layer of brick cause?

f) Find the depth to which the bullet penetrates the brick.

4. The graph represents a two-minute section of a car journey.

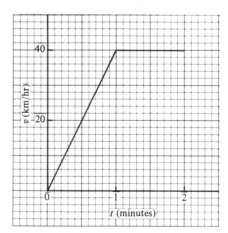

Find

a) the acceleration, in m/s², of the car during the first minute

b) the distance, in metres, travelled by the car during the 2 minutes.

5. A cross-country runner covers three sections of the course in succession. The first is a downhill sweep, then there is a level section followed by a hill climb. The graph below shows the speed, v, of the runner over the three sections.

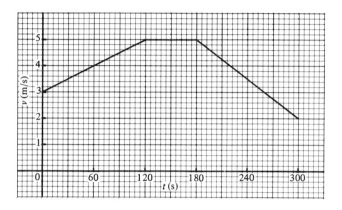

Find

a) the acceleration of the runner on the downhill section

b) the constant speed over the level section

c) the deceleration of the runner during the hill climb

d) the distance covered on the level section

e) the distance covered on the hill climb.

6. The graph represents the journey made by a bus between two bus stops.

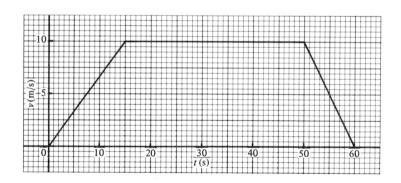

a) What is the acceleration of the bus?

b) What is the deceleration of the bus?

c) What distance does the bus cover while accelerating?

d) What distance does the bus cover while decelerating?

e) What is the distance between the two bus stops?

CURVED VELOCITY-TIME GRAPHS

When acceleration is not constant, the velocity–time graph is a curved line.

For example, suppose that a ball rolls along the ground in such a way that, t seconds after starting, its velocity is v m/s where

$$v = t(10 - t)$$

Forming a table listing values of v for values of t from 0 to 10 gives

t	0	1	2	4	5	6	8	9	10
$(10 - t)$	10	9	8	6	5	4	2	1	0
v	0	9	16	24	25	24	16	9	0

from which we draw the following velocity–time graph.

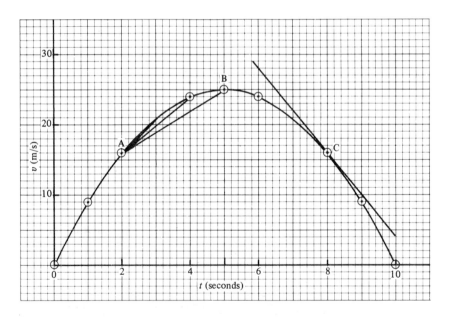

From the graph we see that the speed of the ball increases for 5 seconds, i.e. the ball accelerates for 5 seconds. Then the speed decreases for 5 seconds, i.e. the ball decelerates for 5 seconds.

The maximum speed is 25 m/s and occurs 5 seconds after the start.

The *average acceleration* over an interval of time is the steady acceleration that gives the same final velocity.

In the interval of time from $t = 2$ to $t = 5$, a steady increase in velocity is represented by the straight line AB. Hence the average acceleration over this interval of time is represented by the gradient of the chord AB.

Now consider *acceleration at an instant*. On the graph, the gradient of the tangent at the point C represents the rate at which v is changing at the instant when $t = 8$, i.e. the acceleration at the instant when $t = 8$. Hence

> **acceleration at an instant is represented by the gradient of the tangent to the velocity–time curve at that instant.**

Consider the *distance travelled* in an interval of time. When the acceleration is constant, the velocity–time graph is a straight line and the distance is represented by the area under the straight line. Now a curve can be represented approximately by a succession of short straight lines; the diagram shows an approximation to the curve between A and B.

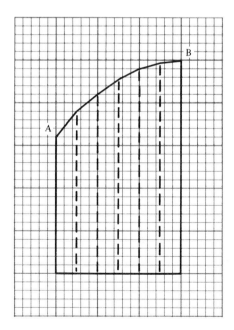

The sum of the areas under each of these straight lines gives a very good approximation for the distance covered in the interval from $t = 2$ to $t = 5$. Hence

> **the distance covered in an interval of time is represented by the area under the velocity–time graph for that interval.**

EXERCISE 18j

The graph represents the first five minutes of the journey of a train.

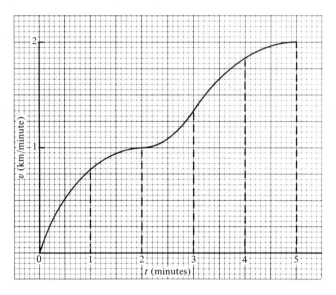

Find   a) the time at which the acceleration is zero
      b) the distance covered by the train in the 5 minutes.

a)   (The tangent has zero gradient where $t = 2$)

The acceleration is zero when $t = 2$

b)   Using six strips each of width 1 unit gives six trapeziums.

Therefore the area is approximately

$$\tfrac{1}{2}(0 + 0.8)(1) + \tfrac{1}{2}(0.8 + 1)(1) + \tfrac{1}{2}(1 + 1.35)(1)$$
$$+ \tfrac{1}{2}(1.35 + 1.85)(1) + \tfrac{1}{2}(1.85 + 2)(1)$$
$$= 0.4 + 0.9 + 1.175 + 1.6 + 1.925 = 6$$

Distance covered ≈ 6 km.

1. The graph illustrates the motion of a ball thrown along the ground.

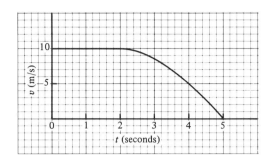

a) Find the velocity of the ball when $t = 4$

b) Is the ball accelerating or decelerating from $t = 2$ to $t = 5$?

c) Find the distance the ball moves before it stops. (Use five strips each of width 1 unit.)

2. The graph illustrates a five-second interval of a car journey.

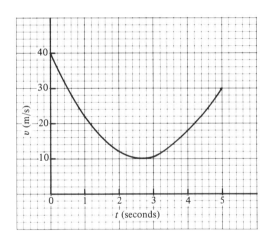

a) Find the distance covered by the car in these five seconds.

b) The following statements about this graph are either true or false. If the statement is true write T, if it is false write F.

i) The car is accelerating when $t = 4$.

ii) The car comes momentarily to rest when $t = 2.5$.

iii) The car changes direction halfway through this interval of time.

iv) The car's speed is decreasing for the first 2 seconds.

3. A rocket is fired and its velocity, v km/minute, t minutes after firing is given by

$$v = t^3$$

Copy and complete the following table.

t	0	1	2	3	4
v	0		8		

Use scales of $2\,\text{cm} \equiv 1$ minute and $1\,\text{cm} \equiv 10\,\text{km/min}$ to draw the velocity–time graph.

From your graph find

a) the acceleration 2 minutes after firing

b) the velocity after $2\frac{1}{2}$ minutes

c) the time when the velocity is $20\,\text{km/min}$

d) the distance covered in the first 3 minutes

e) the distance covered in the fourth minute.

4. The table shows the velocity, v m/s, of a helium filled balloon t seconds after being released in the air on a calm day.

t	0	1	2	3	4	5	6
v	0	5	8	9	8	5	0

Use scales of $1\,\text{cm} \equiv 1$ second and $1\,\text{cm} \equiv 1\,\text{m/s}$ to draw the velocity–time graph.

From your graph find

a) the maximum velocity of the balloon and the time when this occurs

b) the velocity after $1\frac{1}{2}$ s

c) the acceleration 1 second after release

d) the acceleration when $t = 3$

e) the distance covered by the balloon in these 6 seconds.

MIXED EXERCISE

EXERCISE 18k

Each question is followed by several alternative answers. Write down the letter that corresponds to the correct answer.

1.

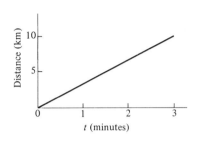

The graph shows an object moving at

A 10 km/min   **B** $3\frac{1}{3}$ km/min   **C** $\frac{3}{10}$ km/min   **D** 30 km/min

2.

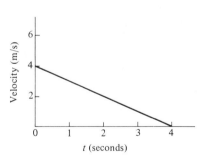

The graph shows an object which in 4 seconds covers a distance of

A 1 m   **B** 8 m   **C** 16 m   **D** −8 m

3. The acceleration of the object whose motion is given in question 2 is

A 1 m/s²   **B** 4 m/s²   **C** 16 m/s²   **D** −1 m/s²

4.

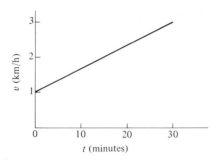

From the graph the distance covered in the thirty minutes is

A 60 km   **B** 1 km   **C** 2 km   **D** 4 km

5.

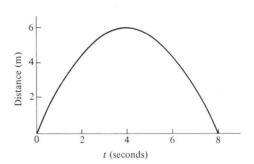

The graph shows an object

A whose velocity is constant

B whose speed when $t = 0$ is zero

C which continues to move away from its initial position

D which returns to its initial position after 8 seconds.

6.

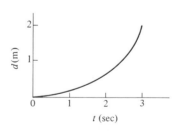

The average velocity from $t = 0$ to $t = 3$ is

A 2 m/s   **B** 1 m/s   **C** $\frac{2}{3}$ m/s   **D** $1\frac{1}{2}$ m/s

7. From the graph in question 6, the acceleration when $t = 2$ could be

A -1 m/s²   **B** zero   **C** 20 m/s²   **D** 1 m/s

8. The distance–time graph representing the motion of a stone dropped from a cliff top could be

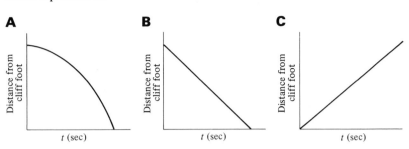

9.

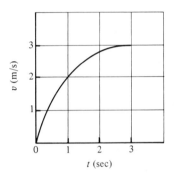

From the graph the distance covered in 3 seconds could be

A 6 m   **B** 8 m   **C** 9 m   **D** 6 km

10. A bead moves along a straight wire with a constant speed for 2 seconds and then its speed decreases at a constant rate to zero. The velocity–time graph illustrating this could be

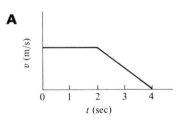

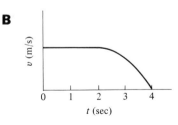

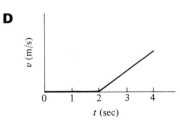

19 VARIATION

RELATIONSHIPS

Frequently, in everyday life, we come across two quantities that appear to be related to each other in some way. The amount I spend on potatoes, when they cost 60 p a bag, depends on the number of bags I buy; the distance I travel in a car, at a constant speed, depends on the length of time that I am travelling; the number of 'singles' I can buy for £10, depends on the price of one record. These are examples of quantities that are related by a simple algebraic equation.

On the other hand, there is no simple algebraic relationship between the amount a person earns and the amount that person spends on food; between our weight and our height; or between how far we travel to school and the time we get up in the morning.

The first exercise in this chapter helps us to recognise some of the simple relationships that connect sets of varying quantities.

EXERCISE 19a

Write down the equation connecting the two variables given in the table.

x	2	3	5	10	12
y	6	9	15	30	36

(We observe that in each case the value of y is three times the value of x)

$$y = 3x$$

In each of the following questions write down the equation connecting the variables.

1.

x	1	2	4	7	10
y	3	6	12	21	30

2.

p	0	1	2	3	4
q	0	1	4	9	16

3.

x	1	2	3	4	5
V	1	8	27	64	125

4.

A	0	4	9	16	25
r	0	2	3	4	5

5.

x	2	4	6	24
y	12	6	4	1

6.

r	0	2	4	6	10
s	0	0.2	0.4	0.6	1

7.

x	−3	−1	0	2	4
y	36	4	0	16	64

8.

p	−9	−6	−3	2	4
q	4	6	12	−18	−9

9. Using squared paper draw six rectangles such that, in each one the length is twice the breadth. Use the lengths given in the table and complete the table to give the area of each rectangle.
What is the equation connecting the area (A) and the length (L)?

Length of rectangle (L) in cm	2	4	5	6	8	10
Area of rectangle (A) in cm^2						

10. Copy the triangle given below on squared paper. Its base is 3 cm and its height is 2 cm.

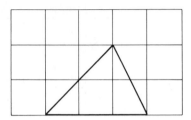

Draw three further similar triangles with bases 6 cm, 9 cm and 12 cm whose heights will be 4 cm, 6 cm and 8 cm respectively. Find the area of each triangle. Use these values to complete the following table.

Base (b) in cm	3	6	9	12
Area (A) in cm^2				

What equation connects A and b?

11. For this question imagine that you have a quantity of identical cubes. Cubes of sugar or Oxo cubes would be suitable. Use these cubes to build bigger cubes whose sides are larger than the basic cube by factors of 2, 3, 4 and 5. It will help if you draw diagrams. How many times larger is the volume of each of these cubes than the volume of the basic cube, i.e. how many of the smallest cubes are required to make each of the larger cubes? Use your results to complete the following table.

Factor by which the side of the basic cube is multiplied (x)	2	3	4	5
Factor by which the volume of the basic cube is multiplied (y)				

The questions in this exercise have illustrated several different ways in which varying quantities may be related. An increase in one quantity may lead to an increase or a decrease in the other.

DIRECT VARIATION

Consider the total cost for a group of people to attend a concert. The varying costs, depending on the size of the group, are given in the table.

Number of people (N)	5	10	15	25	35	50
Total cost in £ (C)	20	40	60	100	140	200

The two quantities C and N are connected by the equation $C = 4N$ i.e. the value of C is always four times the value of N.
This relation is called *direct proportion*.

In general, if two variables Y and X are in direct proportion then $Y = kX$ where k is the constant of proportion.

Sometimes we say that y varies directly with x. This gives exactly the same equation, i.e. $y = kx$ where k is the constant of variation.

In mathematics we are always looking for shorter ways of writing things. Instead of writing the relation 'y is directly proportional to x' or 'y varies directly as x' we sometimes write $y \propto x$, from which we can write the equation $y = kx$ where k is some constant.

EXERCISE 19b

1. Copy and complete the table so that $y \propto x$.

x	2			7	8	
y	20	40				95

What is the equation connecting x and y?

2. Copy and complete the table so that $C \propto r$.

r			3	5		8
C	6	18			36	48

What is the equation connecting C and r?

3. Copy and complete the table so that $C \propto n$.

Number of units of electricity used (n)	100	120	142	260	312	460
Total cost in pence (C)	600		852	1560		2760

What meaning can you give to the constant of proportion?

4. Copy and complete the table so that $Y \propto X$.

Number of oranges bought (X)	2	4	7	9	11	15
Total cost in pence (Y)	20	40		90		150

What meaning can you give to the constant of variation ?

If y varies directly as x and $y = 2$ when $x = 3$, find
a) y when x is 9   b) x when y is 18

$$y \propto x$$

i.e. $y = kx$ where k is a constant

But $y = 2$ when $x = 3$

$\therefore$ $2 = k \times 3$

i.e. $3k = 2$

or $k = \dfrac{2}{3}$

so $y = \dfrac{2}{3}x$

a)   If $x = 9$, $y = \dfrac{2}{3} \times 9$

                    $= 6$

b)   If $y = 18$, $18 = \dfrac{2}{3}x$

i.e. $54 = 2x$

                    $27 = x$

$\therefore$ $x = 27$

---

5. y varies directly as x and $y = 21$ when $x = 7$.
Find   a) y when $x = 3$ b) x when $y = 48$.

6. $y \propto x$ and $y = 6$ when $x = 24$.
Find   a) y when $x = 6$ b) x when $y = 5$.

7. s varies directly as t and $s = 35$ when $t = 5$.
Find   a) s when $t = 3$ b) t when $s = 49$.

8. y varies as $3x - 4$ and $y = 33$ when $x = 5$.
Find   a) y when $x = 2$ b) x when $y = 15$.

9. P is directly proportional to Q and $P = 15$ when $Q = 50$.
Find   a) P when Q is 70   b) Q when P is 12.

10. W is directly proportional to S and $W = 8$ when S is 10.
Find   a) W when S is 30   b) S when W is 12.

11. Y is directly proportional to X and $Y = 45$ when $X = 18$.
Find   a) Y when X is 6   b) X when Y is 20.

DEPENDENT AND INDEPENDENT VARIABLES

Usually one of the quantities varies because of a change in the other. In question 3 above the total cost goes up because the number of units of electricity used goes up, i.e. the variation in the first quantity (C) *depends* on the change in the other (n). C is called a *dependent variable* while n is referred to as an *independent variable*.

Similarly, when the radius of a circle increases, the area of the circle increases. Therefore the radius is the independent variable and the area is the dependent variable.

The dependent variable is sometimes proportional to a power of the independent variable. For example, the area of a circle is directly proportional to the *square* of its radius and we can write $A \propto R^2$ or $A = kR^2$ (k has a special value in this case; can you say what it stands for?)

Similarly, if the safe speed (V) at which a car can round a bend varies as the square root of the radius of the bend (R) then $V \propto \sqrt{R}$ or $V = k\sqrt{R}$.

EXERCISE 19c

1. Copy and complete the table so that $y \propto x^2$

x	0		3	4	5	
y		12	27		75	192

What is the equation connecting x and y?

2. Copy and complete the table so that $s \propto t^2$.

t	2	4		6	10
s		80	125	180	

What is the equation connecting s and t?

3. Copy and complete the table so that $y \propto x^2$

x	−3	−1	0	2	4	7
y				16		196

What is the equation connecting x and y?

If y is directly proportional to the square of x and $y = 3$ when $x = 1$, find

a) y when x is 4   b) x when y is $\frac{3}{4}$.

$$y \propto x^2$$

i.e. $y = kx^2$ where k is a constant

But $y = 3$ when $x = 1$

$\therefore$ $3 = k \times 1^2$

i.e. $k = 3$

so $y = 3x^2$

a)   If $x = 4$, $y = 3 \times 4^2$

$= 3 \times 16$

$= 48$

b)   If $y = \frac{3}{4}$, $\frac{3}{4} = 3x^2$

$x^2 = \frac{1}{4}$ (dividing both sides by 3)

$x = \pm\frac{1}{2}$

4. y is directly proportional to the square of x and $y = 18$ when $x = 3$.
Find   a) y when $x = 4$ b) x when $y = \frac{1}{2}$.

5. y varies as the square of x and $y = 48$ when $x = 4$. Show that $y = 3x^2$
and find   a) y when $x = \frac{1}{2}$ b) x when $y = \frac{1}{3}$.

6. $P \propto Q^2$ and $P = 12$ when $Q = 4$.
Find   a) P when $Q = 12$ b) the positive value of Q when $P = 48$.

If y is directly proportional to the cube of x and $y = 4$ when
$x = 2$ find

a) y when $x = 4$ b) x when $y = \frac{1}{2}$.

$$y \propto x^3$$

i.e. $$y = kx^3$$

But $y = 4$ when $x = 2$

$$\therefore \qquad 4 = k \times 2^3$$

i.e. $$8k = 4$$

$$k = \frac{1}{2}$$

so $$y = \frac{1}{2}x^3$$

a)   If $x = 4$, $y = \frac{1}{2} \times 4^3$

$$= 32$$

b)   If $y = \frac{1}{2}$, $\frac{1}{2} = \frac{1}{2}x^3$

i.e. $$x^3 = 1$$

$$x = 1$$

---

7. Copy and complete the table so that $V \propto H^3$.

H	2			6	8	10
V	2	16	54			

What is the equation connecting V and H?

8. Copy and complete the table so that $y \propto x^3$.

x	3	6	9	12	15
y		72		576	

What is the equation connecting x and y?

9. $y \propto x^3$ and $y = 3$ when $x = 2$.
Find   a) y when $x = 4$ b) x when $y = 81$.

10. If y varies directly as the cube of x and $y = 64$ when $x = 2$,
find   a) y when $x = 3$ b) x when $y = 8$.

11. W is proportional to the cube of H and $W = 32$ when $H = 4$.
Find   a) W when $H = 6$ b) H when $W = 4$.

12. Copy and complete the table so that $V \propto \sqrt{R}$.

R	0	1	4		25
V			8	12	

What is the equation connecting V and R?

13. Y varies directly as the square root of X and $Y = 1$ when $X = 100$.
Find   a) Y when $X = 400$ b) X when $Y = 3$.

14. Plot the graph of y against x for the following data.

x	1	4	9	16	25
y	1	2	3	4	5

Is the graph a straight line? If it is not, complete the following table and plot the graph of y against $\sqrt{x}$.

$\sqrt{x}$					
y	1	2	3	4	5

Is this graph a straight line? What is the equation connecting x and y?

15. Plot the graph of y against x for the following data.

x	1	2	3	4	5
y	0.5	4	13.5	32	62.5

Is the graph a straight line? If not, plot the graphs of y against x^2 and y against x^3. What is the equation connecting x and y?

INVERSE VARIATION

When one quantity decreases as the other increases, the two are said to be inversely proportional, provided that their product always gives the same value. For example, the number of similar postage stamps I can buy for £4.80 depends on their price. At 10p each I can buy 48, at 12p each 40, at 20p each 24 and at 40p each 12. In each case the product of the number of stamps and the price of one stamp is 480p.

EXERCISE 19d

In each question from 1 to 3 complete the given table and show that the product of the varying quantities is constant. Write down the equation connecting these varying quantities.

1.

Cost of a birthday card in pence (C)	25	50	100	125
Number of cards that can be bought for £5 (N)	20		5	

2.

Number of similar magazines a boy could buy with his pocket money (N)	12	9	8	6
Cost of one magazine in pence (C)	60		90	

3.

Pressure (P)	4	5	6	8	12
Volume (V)	30		20		10

In questions 4 to 6 write down the equations connecting x and y.

4.

x	1	2	3	4	6	12
y	12	6	4	3	2	1

5.

x	36	24	18	12	8
y	2	3	4	6	9

6.

x	10	5	1	0.5	0.25
y	0.1	0.2	1	2	4

EQUATIONS FOR INVERSE VARIATION

A teacher has £120 to spend on books. The table shows the number of books of various prices that can be bought.

Cost per book in £ (c)	1	2	3	4	5	8	10
Number of books that can be bought (N)	120	60	40	30	24	15	12

As the cost per book increases the number of books that can be bought decreases. For example, if the cost per book doubles the number of books that can be bought is halved.

We say N varies inversely as C or N is inversely proportional to C, and write this $N \propto \dfrac{1}{C}$ or $N = \dfrac{k}{C}$

In this case $N = \dfrac{120}{C}$ hence $NC = 120$.

This confirms the definition on p 361, that if two quantities are inversely proportional, their product is constant.

i.e. if $y = \dfrac{k}{x}$ then $xy = k$.

Similarly if p is inversely proportional to the square of q then

$$p \propto \frac{1}{q^2}, \quad p = \frac{k}{q^2} \quad \text{and again} \quad pq^2 = k$$

EXERCISE 19e

Copy and complete the table so that $y \propto \dfrac{1}{x^2}$

x	3	5			10	15
y	100	36	25			

If $\qquad\qquad\qquad y = \dfrac{1}{x^2}$

then $\qquad\qquad\qquad y = \dfrac{k}{x^2}$

But $y = 100$ when $x = 3$

$\therefore$
$$100 = \frac{k}{9}$$

i.e. $k = 900$ so $y = \dfrac{900}{x^2}$

(Check: when $x = 5$, $y = \dfrac{900}{25} = 36$)

If $x = 10$, $y = \dfrac{900}{100} = 9$

If $x = 15$, $y = \dfrac{900}{225} = 4$

If $y = 25$, $25 = \dfrac{900}{x^2}$

i.e. $25x^2 = 900$

$$x^2 = \frac{900}{25} = 36$$

$x = 6$ (taking only the positive value)

$\therefore$ the completed table is

x	3	5	6	10	15
y	100	36	25	9	4

1. Copy and complete the table so that $y \propto \dfrac{1}{x}$

x	2	4	6	9	12	
y	18	9			3	2

What is the equation connecting x and y?

2. Copy and complete the table so that $y \propto \dfrac{1}{x^2}$

x	0.5		2	3	6	10
y		36	9			

What is the equation connecting x and y?

3. Copy and complete the table so that $q \propto \dfrac{1}{\sqrt{p}}$

p	0.25		4	9		25
q	120	60	30		15	60

What is the equation connecting p and q?

4. If y is inversely proportional to x, and $y = 8$ when $x = 5$
find   a) y when x is 10   b) x when y is 2   c) y when x is -4.

5. $y \propto \dfrac{1}{\sqrt{x}}$ and $y = 2$ when $x = 4$.
Find   a) y when $x = 9$ b) x when $y = 1$.

6. If p is inversely proportional to v, and $p = 15$ when $v = 20$
find   a) p when $v = 30$ b) v when $p = 7.5$.

7. If P varies inversely as $Q + 2$ and $P = 5$ when $Q = 4$
find   a) P when $Q = 3$ b) Q when $P = 15$.

8. If y is inversely proportional to the square of x and $y = 4$ when $x = 5$
find   a) y when $x = 2$ b) x when $y = 1$.

9. If y varies inversely as x and $y = 6$ when $x = 8$
find   a) y when $x = 12$ b) x when $y = 4$.

10. If y varies inversely as the cube of x and $y = 7$ when $x = 6$
find   a) y when $x = 3$ b) x when $y = 189$.

If y is the constant speed of a train and x is the time it takes to travel a distance k, find the value of n if x and y are related by a law of the form $y \propto x^n$.

Since   distance travelled $=$ constant speed $\times$ time

$$k = y \times x$$

i.e. $$xy = k$$

and $$y = \frac{k}{x}$$

or $$y = kx^{-1}$$

$\therefore$ $$y \propto x^n \quad \text{where} \quad n = -1$$

11. In each of the following cases, x and y are related by a law of the form $y \propto x^n$. Find the value of n.

a) y is the area of a square and x is the length of one side.

b) y is the area of a circle and x is its radius.

c) y is the volume of a sphere and x is its radius.

d) y is the length of a rectangle of constant area and x is its breadth.

e) y is the radius of a circle and x is its area.

f) y is the length of a line in centimetres and x is its length in millimetres.

MIXED EXAMPLES

EXERCISE 19f

1. p varies directly as the square of q and $p = 9$ when $q = 6$.

Find p when q is   a) 2   b) –2   c) 5.

2. A is directly proportional to L and $A = 28$ when $L = 4$.

Find   a) A when $L = 3$ b) L when $A = 42$.

3. $y \propto x^3$ and $y = 48$ when $x = 4$.

Find   a) the formula for y in terms of x b) y when $x = 2$

c) x when $y = 6$.

4. y varies inversely as x and $y = 7$ when $x = 6$.

Find   a) y when $x = 3$ b) x when $y = 14$.

5. y is inversely proportional to x^2 and $y = 4.5$ when $x = 4$.

Find   a) y when $x = 3$ b) x when $y = 8$.

6. Copy and complete the table so that $y \propto x^2$.

x	0	1		4	8
y		0.25	1		16

7. Copy and complete the table so that $t \propto \sqrt{s}$.

s		4	9		
t	0	0.5		1	2

8. Given that y varies as x^n, write down the value of n in each of the following cases:

a) y is the area of a square of side x

b) y is the volume of a cube of side x

c) y is the volume of a cylinder with constant base area A and height x

d) y and x are the sides of a rectangle with a given area.

GRAPHS FROM EXPERIMENTAL DATA

In science we often conduct experiments where one quantity (say q) is measured for various values of another quantity (say p).

If these values are plotted on a graph, the points often lie, more or less, on a straight line.

Consider the following experimental data for two quantities p and q.

p	1	2	3	4	5	6	7
q	4	6.5	7.5	10.5	12	10	16.5

Plotting these points gives this diagram.

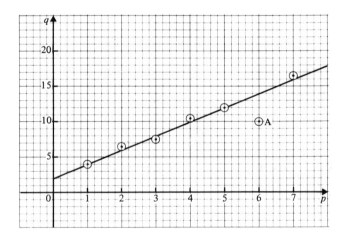

If we assume that the point marked A is the result of an error in measurement and hence ignore it, then the other points can be seen to lie roughly on a straight line. It is unrealistic to expect the points to lie *exactly* on a straight line as measurements cannot be taken so accurately.

Assuming that some measurements are high and some are low, the line that best fits these points can be drawn by positioning the line so that the sum of the distances of points above the line is approximately equal to the sum of the distances of points below the line. This is best judged by eye, using a transparent ruler.

We know that when using x and y axes, the equation of a straight line can be written as $y = mx + c$ where m is the gradient and c is the intercept on the y-axis. In the same way the equation of our line can be written in the form $q = mp + c$ and this gives the relationship between p and q.

From the graph, the gradient of the line is 2 and the intercept on the q-axis is 2.

Therefore the equation of the line is

$$q = 2p + 2$$

i.e. the relationship between p and q is $q = 2p + 2$.

EXERCISE 19g

1. In a spring stretching experiment the following results were obtained.

Stretching force in newtons (F)	0	1	2	3	4	5	6
Extension of spring in mm (E)	0	8	17	25	33	42	50

Use scales of 4 cm to 10 mm and 2 cm to 1 newton to plot a graph of E against F and use it fo find

a) the stretching force required to produce an extension of 20 mm

b) the relationship between E and F.

2. The table shows the results of measuring the length of a column of air at various temperatures.

Temperature in °C (T)	20°	30°	40°	50°	60°	70°	80°
Length of column in cm (l)	7	7.2	7.3	7.6	7.9	8	8.1

Use a scale of 2 cm to 10°C and 2 cm to 1 cm and plot these results. Draw the straight line that best fits these points and use it to find the temperature when the column of air is 7.5 cm long.

3. The relationship between the current, I, flowing through a tangent galvanometer is given by the formula $I = k \tan \theta$ where θ is the angle of deflection. The table shows the results obtained in an experiment.

Current in amps (I)	0.42	0.65	0.89	1.18	1.70	1.88
Mean deflection (θ)	27.6°	38.7°	43°	56.2°	64.4°	67.3°
Tan θ						

a) Copy the table and complete the last row.

b) Plot values of I against values of $\tan \theta$ using a scale of 1 cm to 0.1 units on both axes.

c) By drawing the line that best fits these points, find the value of k.

4. An athlete trained by running round a track. His coach recorded the distance covered by the athlete at various times and the results are shown in the table.

Time in seconds from start (t)	10	20	30	50	80	100
Distance from starting point in metres (d)	42	83	120	210	350	390

Use scales of 4 cm to 100 m and 1 cm to 10 seconds and plot these results on a graph.

a) If the coach made an error on one reading, which one is it?

b) Ignoring the one reading that is wrong, draw the line that best fits the other results.

c) Show that the distance varies directly as the time and hence find the speed of the athlete.

5. The magnification, M, produced by a convex lens of focal length f when the distance of the lens from the image is v is given by the equation

$$M = \frac{v}{f} - 1$$

In an experiment the following values of v and M were obtained

v (cm)	20.3	24.8	27.1	30.4	38.0	42.3	47.4
M	1	1.5	1.75	2.1	2.4	3.25	3.75

Plot a graph of M against v and draw the straight line that best fits this data. Use your graph to find the value of f.

PROBLEMS ON DIRECT AND INVERSE VARIATION ⎯⎯⎯⎯⎯⎯

EXERCISE 19h

A stone falls from rest down a mine shaft. It falls D metres in T seconds where D varies as the square of T. If it falls 20 m in the first 2 s and takes 5 s to reach the bottom, how deep is the shaft ?

$$D \propto T^2$$

i.e. $$D = kT^2$$

But $D = 20$ when $T = 2$

$\therefore$ $$20 = k \times 2^2$$

i.e. $$4k = 20$$

$$k = 5$$

$\therefore$ $$D = 5T^2$$

If $T = 5$, $D = 5 \times 5^2$

$$= 125$$

Therefore the shaft is 125 metres deep.

1. The mass, M kg, of a circular disc of constant thickness varies as the square of its radius, R cm. If a disc of 5 cm radius has a mass of 1 kilogram find

a) the mass of a disc of radius 10 cm

b) the radius of a disc of mass 25 kg.

2. The safe speed, V km/h, at which a car can round a bend of radius R metres varies as $\sqrt{R}$. If the safe speed on a curve of radius 25 m is 40 km/h, find the radius of the curve for which the safe speed is 64 km/h.

3. The time of swing, T s, of a simple pendulum is directly proportional to the square root of its length, L cm. If $T = 2$ when $L = 100$ find
a) T when $L = 64$ b) L when $T = 1\frac{1}{2}$.

4. The extension, x cm, of an elastic string varies as the force, F newtons, used to extend it. If a force of 4 newtons gives an extension of 10 cm find

a) the extension given by a force of 10 newtons

b) the force required to give an extension of 12 cm.

5. The cost of buying a rectangular carpet is directly proportional to the square of its longer side. If a carpet whose longest side is 3 m costs £180 find

a) the cost of a carpet with a longer side of 4 m

b) the length of the longer side of a carpet costing £405.

6. The radius of a circle, r cm, varies as the square root of its area, A cm^2. How does the radius change if the area is increased by

a) a factor of 4   b) a factor of 25 ?

7. Mathematically similar jugs have capacities that vary as the cubes of their heights. If a jug 10 cm high holds $\frac{1}{8}$ litre find

a) the capacity of a jug that is 12 cm high

b) the height of a jug that will hold 1 litre.

8. For a given mass of gas at a given temperature the pressure p varies inversely as the volume, v. If $p = 100$ when $v = 2.4$ find

a) v when $p = 80$ b) p when $v = 2$.

9. For a vehicle travelling between two motorway service stations the time taken is inversely proportional to its speed. If it takes $2\frac{1}{2}$ hours when its speed is 48 m.p.h. find

a) its average speed if it takes 3 hours

b) by how much its average speed must increase if the journey time is to be reduced to 2 hours.

10. By what factor does y change when x is doubled if

a) $y \propto x$ b) $y \propto \dfrac{1}{x}$ c) $y \propto x^2$ d) $y \propto x^3$

11. State the percentage change in y when x is increased by 25% if

a) $y \propto x$ b) $y \propto \dfrac{1}{x}$ c) $y \propto x^2$

20 GENERAL REVISION EXERCISES

1. A motorcyclist buys 20 litres of petrol at 48.7 p per litre.
   a) Find the cost of the petrol bought.
   b) Using $4\ell \approx 7$ pints, find the number of gallons bought.

2. Factorise:
   a) $2x^2 - 6x$ b) $x(x-1) - 3(x-1)$ c) $x^2 - 3x - 10$

3. Solve the equations   a) $\dfrac{3}{x} + \dfrac{1}{2} = 4$
                                b) $x^2 - 5x + 6 = 0$
                                c) $x + y = 2$
                                   $x - y = 4$

4. Simplify   a) $3(2-x) - 2(x-1)$
                    b) $\left(1\frac{1}{2}\right)^{-2}$
                    c) $\dfrac{x-1}{3} - \dfrac{x-4}{6}$

5. A house was bought for £20000 on 1st January 1980. Each year the value of the house appreciates by 10% of its value at the beginning of the year. Find the value of the house on 1st January 1986.

6. Alison is twice as good as Paul at chess. They play two games. What is the probability that
   a) Alison wins both games   b) she wins only one game?

7. A and B are two points 6 cm apart. Sketch the loci that will enable you to shade the region in which every point satisfies both the following conditions.
   a) It is within 4 cm of A.
   b) It is further away from A than it is from B.

8. y varies inversely as x and $y = 8$ when $x = 18$.
   Find a formula for y in terms of x and use it to find
   a) the value of y when $x = 4$ b) the value of x when $y = 9$.

371

9. A ship A is 8 miles due East of a ship B. A third ship C is observed simultaneously from A and B and found to be on a bearing of 290° from A and on a bearing of 020° from B. Draw a sketch showing the positions of A, B and C. Calculate the distances of C from A and from B giving your answers correct to three significant figures.

10.

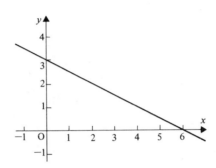

Use the diagram to find

a) the gradient of the line

b) the equation of the line

c) the area enclosed by the line, the x-axis and the ordinates $x = 2$ and $x = 4$.

EXERCISE 20b

1. An automatic cooker works on a 24-hr digital clock. To make the oven cook automatically, both the cooking time and the finishing time have to be programmed in. A casserole which needs $3\frac{1}{2}$ hours cooking-time, has to be ready at 6.30 p.m. and quarter of an hour is required for the oven to heat up. Find

a) the finishing time to be entered on the programme

b) the reading on the clock when the oven comes on.

2. A ball is thrown vertically upward from a point A with speed u m/s. Its distance, s metres, from A, t seconds after being thrown, is given by

$$s = ut - 5t^2$$

a) Find s when $u = 40$ and $t = 5$.

b) Find t when $s = 1$ and $u = 6$.

c) Interpret your answer to (b).

d) Make u the subject of the formula.

3. Simplify: a) $5x \times 3x^2$

   b) $\dfrac{x+1}{2} - \dfrac{x-1}{5}$

   c) $(125)^{1/3}$

4. Use squared paper and the same scale on both axes to draw a *sketch* of the graph of $y = x^2$. On the same axes sketch the graph of $y = x + 1$.

a) Use your sketch to estimate the solution of the equation $x^2 = x + 1$.

b) Calculate the solution of the equation $x^2 = x + 1$, giving your answers to 2 decimal places.

5. The point (x, y) is mapped to the point $(2, 4)$ by the transformation matrix $\begin{pmatrix} 1 & 2 \\ 0 & 1 \end{pmatrix}$. Find x and y.

6.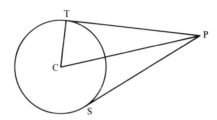

PS and PT are the two tangents from the point P to the circle centre C. P is 20 cm from the centre of the circle and the radius of the circle is 6 cm. Find

a) the lengths of the tangents PT and PS

b) the angle between the tangents.

7. If $n(\mathscr{E}) = 40$, $n(A) = 21$, $n(B) = 14$ and $n(A \cap B) = 5$ draw a Venn diagram and find

a) $n(A \cup B)$ b) $n(A \cap B')$ c) $n(A' \cap B)$ d) $n(A \cup B)'$

8. A racing cyclist did a trial lap of a circuit. His speed was recorded from the start at 5-second intervals of time and the results are given in the table.

Time from start, t sec	0	5	10	15	20	25
Speed, v m/s	0	6	8.2	9.3	9.9	10

Using scales of 2 cm for 5 seconds and 1 cm for 1 m/s draw a velocity–time graph showing this information. Use your graph to find, approximately

a) the speed when $t = 8$

b) the acceleration when $t = 10$

c) the distance covered in the first ten seconds.

9. Two model cars are mathematically similar. One is 8 cm long and the other is 6 cm long.

a) Calculate the width of the smaller car given that the width of the larger car is 4.8 cm.

b) If 13.5 cm³ of metal were used to make the smaller car calculate the volume of metal used to make the larger car.

10.

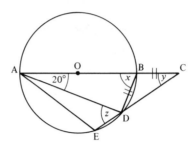

In the diagram, O is the centre of the circle, and $\widehat{EDC}$ is 180°. Find the size of each marked angle, giving reasons for your working.

EXERCISE 20c

1. A car travels from town A to town B at an average speed of 64 km/h in 1 hour 45 minutes.

a) Find the distance covered by the car.

b) Using 5 miles ≈ 8 kilometres, find the distance covered in miles.

2. Solve the equations   a) $5x = 3(8 - x)$

                      b) $2x - y = 5$

                         $x + 2y = 7$

3. A man bought x packets of butter at 50 p each and y packets of tea at 30 p each. Write down an algebraic statement describing each of the following.

a) The cost of the man's purchases was £3.20.

b) The man bought 8 packets altogether.

c) The number of packets bought was less than ten.

d) The cost of the packets was more than £5.

4. Find the value of x if   a) $x^3 = 125$ b) $2^x = \frac{1}{8}$ c) $3^x = 27$

5. Solve the equation $2x^2 - 8x + 3 = 0$ giving your answers correct to two decimal places.

6. A survey of 72 houses showed that 40 houses were owner-occupied, 5 were empty and the remainder were rented. Show this information on a pie chart, giving the angle at the centre of each slice.

7. Draw four copies of the following diagram and use them to shade the regions illustrating the given sets.

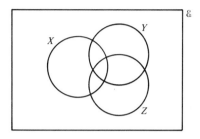

a) $X \cap Y \cap Z$ b) $X' \cap Y$ c) $(X \cup Z)'$ d) $(X \cup Y) \cap Z'$

8. There are 25 pupils in a class and the ratio of girls to boys is $3:2$. What is the probability that a pupil, chosen at random for a place on a quiz team, is a girl?

What is the probability that the next pupil chosen is also a girl?

9.

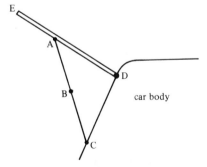

car body

The diagram shows a section through the tailgate of a car, with the tailgate DE in its fully open position. DE is kept in this position by the stay ABC. The stay consists of two rigid bars, AB and BC which are freely hinged together at B. One end of the stay is hinged to the car body at C and the other end of the stay is hinged to the tailgate at A.

With the tailgate fully open, ABC is a straight line, 50 cm long. AD and DC are each 30 cm long.

Using a scale of 1 cm to 5 cm make a scale drawing of this section.

The stay is released by pulling the hinge B away from D. By considering the tailgate closing 20° at a time, or otherwise, draw the path traced out by the hinge B as the door closes.

10. ABCD is a rectangle with A(1, 4), B(4, 4), C(4, 8) and D(1, 8).
S is the transformation 'reflection in the line $y = x$'
T is the transformation 'clockwise rotation of 90° about O'.
Draw x and y axes on squared paper for $-2 \leqslant x \leqslant 10$, $-10 \leqslant y \leqslant 10$.

a) Draw ABCD and its image A'B'C'D' under the transformation S.

b) Draw the image of A'B'C'D' under the transformation T.
Label the image A"B"C"D".

c) Describe the transformation that maps ABCD directly to A"B"C"D".

EXERCISE 20d

1. a) Calculate, exactly, the value of $\dfrac{2.19 - 0.6753}{101 + 889}$

b) Give your answer to (a)   i) in standard form   ii) correct to 1 s.f.

2. Calculate   a) $2\begin{pmatrix} 1 & 7 \\ -2 & 4 \end{pmatrix} - \begin{pmatrix} 3 & -2 \\ 1 & 5 \end{pmatrix}$

b) $\begin{pmatrix} 1 & 7 \\ -2 & 4 \end{pmatrix}\begin{pmatrix} 3 & -2 \\ 1 & 5 \end{pmatrix}$

c) Find the determinant of $\begin{pmatrix} 1 & 7 \\ -2 & 4 \end{pmatrix}$

3. A class of 30 pupils was given a test, marked out of 10. The distribution of the marks is given in the table.

Mark	0	1	2	3	4	5	6	7	8	9	10
Frequency	1	1	2	1	0	4	4	5	6	4	2

Find   a) the modal mark   b) the mean mark   c) the median mark.

4. From a point A on level ground, 30 m from the foot, B, of a tower, the angle of elevation of the top, C, of the tower is 54°. Calculate the height of the tower.

5. a) Simplify $\dfrac{5}{t+3} + \dfrac{2}{t-2}$

b) Solve the equation $5x^2 - 4x - 8 = 0$ giving your answers correct to two decimal places.

6. a) The size of each exterior angle of a regular polygon is $x°$ and the size of each interior angle is $2x°$. Find x and name the polygon.

b) In $\triangle ABC$, $\widehat{A} = 90°$, AB $= 15$ cm and AC $= 9$ cm.
Find BC and angle ACB.

7.

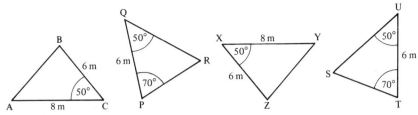

Find two pairs of congruent triangles, giving reasons for your answers.

8. Neil has twenty-two coins in his pocket and they have a total value of £8. The coins are of value £1, 50 p and 5 p. There are three times as many 50 p coins as there are £1 coins.
Let the number of £1 coins be x and write down the number of 5 p coins.

Find the value (in pence) in terms of x of

a) the £1 coins b) the 50 p coins c) the 5 p coins.

Form an equation in x and solve it to find the number of each kind of coin in Neil's pocket.

9.

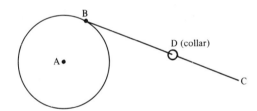

A wheel of diameter 4 cm can rotate about its centre A. A rod BC which is 8 cm long is freely pivoted to the rim of the wheel at B and passes through a fixed collar D which is 5 cm from A (i.e. AD = 5 cm).
Draw, full size, the diagram when A, B, D and C are in one straight line, in that order. By considering different positions of B (let AB turn through 30° clockwise for each new position), draw the path traced out by C for each complete turn of the wheel.

10. Copy and complete the table so that $R \propto \sqrt{A}$

A	1	4	9		
R		0.4		0.8	1

What formula connects R and A? Use this formula to find A when $R = 2$.

EXERCISE 20e

1. A radio costs £55.66 including VAT at 15%. Find the cost of the radio without VAT.

2. Factorise   a) $x^2 - 9$ b) $3xy - 6x^2$ c) $2x^2 - 4x - 6$.

3. A delivery company offers 'same day' delivery in the UK on parcels and packets at the following rates:

£10 for a packet weighing up to 3 kg
and then 50p per 500 g or part of 500 g.

Find the charge for delivering a packet weighing

a) 2.7 kg   b) 5.2 kg   c) 4.8 kg.

4. Given that $a = \dfrac{1}{b} + \dfrac{1}{c}$ find

a) the value of a when $b = 3$ and $c = -4$
b) the value of b when $a = 2$ and $c = 5$.
c) Make c the subject of the formula.

5.

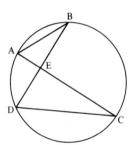

A, B, C and D are four points on the circumference of a circle. Prove that △ABE is similar to △DCE.

6. a) On a map whose scale is $1:5000$, the area of a field is $4\,\text{cm}^2$. Find the area of the actual field.

b)

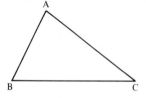

Triangles ABC and DEF are similar and the scale factor for reducing △ABC to △DEF is $\frac{1}{2}$.

i)   If AB = 8 cm, find DE.
ii) If BC = 13 cm, find EF.
iii) If DF = 4.5 cm, find AC.
iv) If area △ABC = 35.5 cm² find area △DEF.

7. The sum of the squares of two consecutive even numbers, the first of which is x, is 164. Form an equation in x and solve it to find these numbers.

8.

Paddington*	d	00 50	06 55	08 00	09 00	09 35	10 00	11 00	11 35	12 00	13 00	14 00	15 00
Slough	d	___	07 08	___	___	09 48	___	___	11 48	___	___	___	___
Reading A*	d	___	07 23	08 24	09 24	10 03	___	11 24	12 03	___	13 24	___	15 24
Didcot	d	___	07 34	___	___	10 16	___	___	12 16	___	___	___	___
Swindon	d	02 17	07 54	___	___	10 37	___	___	12 37	___	___	14 49	___
Bristol Parkway	d	___	08 21	09 12	10 12	11 01	___	12 12	13 01	___	14 12	15 14	16 12
Newport	a	03 10	08 42	09 32	10 32	11 21	11 27	12 32	13 21	13 28	14 32	15 35	16 32
Cardiff Central	a	03 39	05 58	09 48	10 48	11 39	11 43	12 48	13 39	13 44	14 48	15 51	16 48

Paddington*	d	16 00	16 25	17 00	18 00	18 17
Slough	d	___				
Reading A*	d	___	16 49	___	18 25	___
Didcot	d	___				
Swindon	d	___	17 17	___	___	___
Bristol Parkway	d	17 08	17 43	18 09	19 12	19 46
Newport	a	17 28	18 04	18 32	19 32	20 08
Cardiff Central	a	17 44	18 22	18 46	19 48	20 26

Use this extract from the timetable for trains from London (Paddington) to Cardiff to find

a) the time of departure of the latest train from London to arrive in Cardiff before 4 p.m.

b) the time of departure of the fastest train and its journey-time

c) the time of departure of the slowest train and its journey-time.

9. The histogram shows the distribution of marks obtained in a test by a group of pupils.

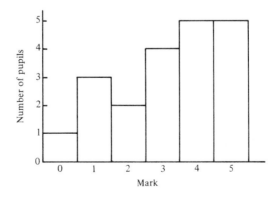

State whether each of the following statements is true or false

a) Twenty pupils took the test.

b) The median mark is $2\frac{1}{2}$.

c) More than half the pupils got a mark of 3 or more.

d) The mean mark is 3.2.

10. a) A rectangular sheet of card, 10 cm by 30 cm, is curved round to form a cylinder 10 cm high. Assuming that the straight edges of the card just meet and there is no overlap, find

   i) the radius of the cylinder

   ii) the volume enclosed by the cylinder.

b)

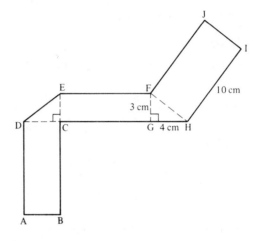

This net makes a triangular prism when folded along the dotted lines.

Find   i) which letter is joined to B

        ii) which letter is joined to J

        iii) the volume of the prism.